The q,t-Catalan Numbers and the Space of Diagonal Harmonics

With an Appendix on the Combinatorics of Macdonald Polynomials

University LECTURE Series

Volume 41

The q,t-Catalan Numbers and the Space of Diagonal Harmonics

With an Appendix on the Combinatorics of Macdonald Polynomials

James Haglund

American Mathematical Society
Providence, Rhode Island

2000 *Mathematics Subject Classification.* Primary 05E05, 05A30; Secondary 05A05.

For additional information and updates on this book, visit
www.ams.org/bookpages/ulect-41

Library of Congress Cataloging-in-Publication Data

Haglund, James.
    The q, t-Catalan numbers and the space of diagonal harmonics : with an appendix on the combinatorics of Macdonald polynomials / James Haglund.
        p. cm. — (University lecture series, ISSN 1047-3998 ; v. 41)
    Includes bibliographical references.
    ISBN-13: 978-0-8218-4411-3 (alk. paper)
    ISBN-10: 0-8218-4411-3 (alk. paper)
    1. Symmetric functions.   2. Combinatorial analysis.   3. Polynomials.   4. Harmonic functions.
I. Title.

QA212.H34   2008
515′.22—dc22

2007060570

Contents

Preface

This book is an expanded version of lecture notes for a topics course given by the author at the University of Pennsylvania during the spring of 2004 on the combinatorics of the q, t-Catalan numbers and the space of diagonal harmonics. These subjects are closely related to the study of Macdonald polynomials, which are an important family of multivariable orthogonal polynomials introduced by Macdonald in 1988 with applications to a wide variety of subjects including Hilbert schemes, harmonic analysis, representation theory, mathematical physics, and algebraic combinatorics. Many wonderful results about these polynomials from analytic, algebraic, and geometric viewpoints have been obtained, but the combinatorics behind them had remained rather impenetrable. Toward the end of the spring 2004 semester the author, inspired primarily by new combinatorial identities involving diagonal harmonics discussed in Chapter 6 of this book, was led to a combinatorial formula for Macdonald polynomials. The discovery of this formula, which was proved in subsequent joint work with Mark Haiman and Nick Loehr, has resulted in a volume of broader interest, as in Appendix A we include a discussion of the formula, its proof, and the nice applications it has to the theory of symmetric functions.

Among these applications we might draw the reader's attention to the short, elegant proof in Appendix A of Lascoux and Schützenberger's "cocharge" theorem on Hall-Littlewood polynomials, a fundamental result in the theory of symmetric functions whose original proof was neither short nor elegant. Another application of the combinatorial formula is a way of writing the Macdonald polynomial as a positive sum of LLT polynomials, which are symmetric functions introduced by Lascoux, Leclerc, and Thibon. This decomposition is especially significant in view of two recent preprints, one by Grojnowski and Haiman and another by Sami Assaf, which contain proofs that the coefficients of LLT polynomials, when expanded in terms of Schur functions, are positive. Although Grojnowski and Haiman's proof uses Kazhdan-Lusztig theory and algebraic geometry, Assaf's proof is a self-contained 21 page combinatorial argument. Thus we now have an accessible, combinatorial proof that Macdonald polynomials are Schur positive. (This Macdonald positivity result was first proved in 2000 by Haiman using properties of the Hilbert scheme from algebraic geometry.) The attempt to understand the combinatorics of Macdonald polynomials is what led Garsia and Haiman to study diagonal harmonics and has been the motivation behind quite a bit of research in algebraic combinatorics over the last 20 years.

Chapter 1 contains some well-known introductory material on q-analogues and symmetric functions. Chapter 2 gives some of the historical background and basic theorems involving Macdonald polynomials and diagonal harmonics, including a discussion of how a certain S_n action on the space of diagonal harmonics leads to a number of beautiful and deep combinatorial problems. Chapters $3 - 6$ deal with the combinatorics of the character induced by this action. The most fundamental object in this subject is the q, t-Catalan numbers, the focus of Chapter 3. From there we move on to a study of the q, t-Schröder numbers in Chapter 4, which are a bigraded version of the multiplicity of a hook shape in the character. Chapter 5 deals with a (conjectured) expression for the bigraded Hilbert series, which has an elegant expression in terms of combinatorial objects called parking functions. In Chapter 6 we study the "shuffle conjecture" of Haiman, Loehr, Remmel, Ulyanov, and the author which gives a combinatorial prediction, parameterized in terms of parking functions, for the expansion of the character into monomials. This conjecture includes all of the results and conjectures from Chapters $3 - 5$ as special cases. Chapter 7 consists of an exposition of the proof of the broadest special case of this conjecture that we can prove, that of hook shapes. The proof involves the manipulation of technical symmetric function identities involving plethysm and Macdonald polynomials. These identities are rather difficult to learn about from reading journal articles, and it is hoped this chapter will be a useful guide to readers interested in learning these subjects. Appendix B contains a discussion of an amazing extension of the shuffle conjecture recently proposed by Loehr and Warrington.

There are homework exercises interspersed throughout the text, in strategically chosen locations, to help the reader absorb the material. Solutions to all the exercises are given in Appendix C. The book is meant to have value either as a text for a topics course in algebraic combinatorics, a guide for self-study, or a reference book for researchers in this area.

The author would like to thank Mahir Can, Nick Loehr, Sarah Mason, Jaren Smith, Chunwei Song, Greg Warrington, and the other students and postdocs attending the course at Penn in the spring of 2004 for many helpful comments on the exposition of the material. In addition he would like to thank Laura Stevens and Sami Assaf for reading through the text and providing a list of errata. The author is also thankful for the support and encouragement of Edward Dunne of the AMS editorial staff, who first suggested the author write this book after hearing an address by the author on this subject at the AMS sectional meeting in Orlando in November 2002. During the course of working on this book, the author was supported by NSA Grant #02G-193 and NSF Grant DMS 0553619.

James Haglund

CHAPTER 1

Introduction to q-Analogues and Symmetric Functions

Permutation Statistics and Gaussian Polynomials

In combinatorics a *q-analogue* of a counting function is typically a polynomial in q which reduces to the function in question when $q = 1$, and furthermore satisfies versions of some or all of the algebraic properties, such as recursions, of the function. We sometimes regard q as a real parameter satisfying $0 < q < 1$. We define the q-analogue of the real number x, denoted $[x]$ as

$$[x] = \frac{1 - q^x}{1 - q}.$$

By l'Hôpital's rule, $[x] \to x$ as $q \to 1^-$. Let $\mathbb{N}$ denote the nonnegative integers. For $n \in \mathbb{N}$, we define the q-analogue of $n!$, denoted $[n]!$ as

$$[n]! = \prod_{i=1}^{n}[i] = (1 + q)(1 + q + q^2) \cdots (1 + q + \ldots + q^{n-1}).$$

We let $|S|$ denote the cardinality of a finite set S¿ By a *statistic* on a set S we mean a combinatorial rule which associates an element of $\mathbb{N}$ to each element of S. A permutation statistic is a statistic on the symmetric group S_n. We use the one-line notation $\sigma_1\sigma_2\cdots\sigma_n$ for the element $\sigma = \begin{pmatrix} 1 & 2 & \ldots & n \\ \sigma_1 & \sigma_2 & \ldots & \sigma_n \end{pmatrix}$ of S_n. More generally, a word (or multiset permutation) $\sigma_1\sigma_2\cdots\sigma_n$ is a linear list of the elements of some multiset of nonnegative integers. (The reader may wish to consult [**Sta86**, Chapter 1] for more background on multiset permutations.) An *inversion* of a word σ is a pair (i, j), $1 \le i < j \le n$ such that $\sigma_i > \sigma_j$. A *descent* of σ is an integer i, $1 \le i \le n - 1$, for which $\sigma_i > \sigma_{i+1}$. The set of such i is known as the descent set, denoted $\text{Des}(\sigma)$. We define the inversion statistic $\text{inv}(\sigma)$ as the number of inversions of σ and the major index statistic $\text{maj}(\sigma)$ as the sum of the descents of σ, i.e.

$$\text{inv}(\sigma) = \sum_{\substack{i<j \\ \sigma_i > \sigma_j}} 1, \quad \text{maj}(\sigma) = \sum_{\substack{i \\ \sigma_i > \sigma_{i+1}}} i.$$

A permutation statistic is said to be *Mahonian* if its distribution over S_n is $[n]!$.

THEOREM 1.1. *Both inv and maj are Mahonian, i.e.*

(1.1) $$\sum_{\sigma \in S_n} q^{inv(\sigma)} = [n]! = \sum_{\sigma \in S_n} q^{maj(\sigma)}.$$

PROOF. Given $\beta \in S_{n-1}$, let $\beta(k)$ denote the permutation in S_n obtained by inserting n between the $(k-1)$st and kth elements of β. For example, $2143(3) =$

1

21543. Clearly $\mathrm{inv}(\beta(k)) = \mathrm{inv}(\beta) + n - k$, so

$$(1.2) \qquad \sum_{\sigma \in S_n} q^{\mathrm{inv}(\sigma)} = \sum_{\beta \in S_{n-1}} (1 + q + q^2 + \ldots + q^{n-1}) q^{\mathrm{inv}(\beta)}$$

and thus by induction inv is Mahonian.

A modified version of this idea works for maj. Say the descents of $\beta \in S_{n-1}$ are at places $i_1 < i_2 < \cdots < i_s$. Then

$$\mathrm{maj}(\beta(n)) = \mathrm{maj}(\beta), \quad \mathrm{maj}(\beta(i_s + 1)) = \mathrm{maj}(\beta) + 1,$$
$$\ldots, \mathrm{maj}(\beta(i_1 + 1)) = \mathrm{maj}(\beta) + s, \quad \mathrm{maj}(\beta(1)) = s + 1.$$

If the non-descents less than $n - 1$ of β are at places $\alpha_1 < \alpha_2 < \cdots < \alpha_{n-2-s}$, then

$$\mathrm{maj}(\beta(\alpha_1 + 1)) = \mathrm{maj}(\beta) + s - (\alpha_1 - 1) + \alpha_1 + 1 = \mathrm{maj}(\beta) + s + 2.$$

To see why, note that $s - (\alpha_1 - 1)$ is the number of descents of β to the right of α_1, each of which will be shifted one place to the right by the insertion of n just after β_{α_1}. Also, we have a new descent at $\alpha_1 + 1$. By similar reasoning,

$$\mathrm{maj}(\beta(\alpha_2)) = \mathrm{maj}(\beta) + s - (\alpha_2 - 2) + \alpha_2 + 1 = \mathrm{maj}(\beta) + s + 3,$$

$$\vdots$$

$$\mathrm{maj}(\beta(\alpha_{n-2-s})) = \mathrm{maj}(\beta) + s - (\alpha_{n-2-s} - n - 2 - s) + \alpha_{n-2-s} + 1$$
$$= \mathrm{maj}(\beta) + n - 1.$$

Thus

$$(1.3) \qquad \sum_{\sigma \in S_n} q^{\mathrm{maj}(\sigma)} = \sum_{\beta \in S_{n-1}} (1 + q + \ldots + q^s + q^{s+1} + \ldots + q^{n-1}) q^{\mathrm{maj}(\beta)}$$

and again by induction maj is Mahonian. □

Major P. MacMahon introduced the major-index statistic and proved it is Mahonian [**Mac60**]. Foata [**Foa68**] found a map ϕ which sends a permutation with a given major index to another with the same value for inv. Furthermore, if we denote the descent set of σ^{-1} by $\mathrm{Ides}(\sigma)$, then ϕ fixes $\mathrm{Ides}(\sigma)$. The map ϕ can be described as follows. If $n \le 2$, $\phi(\sigma) = \sigma$. If $n > 2$, we add a number to ϕ one at a time; begin by setting $\phi^{(1)} = \sigma_1$, $\phi^{(2)} = \sigma_1 \sigma_2$ and $\phi^{(3)} = \sigma_1 \sigma_2 \sigma_3$. Then if $\sigma_3 > \sigma_2$, draw a bar after each element of $\phi^{(3)}$ which is greater than σ_3, while if $\sigma_3 < \sigma_2$, draw a bar after each element of $\phi^{(3)}$ which is less than σ_3. Also add a bar before $\phi_1^{(3)}$. For example, if $\sigma = 4137562$ we now have $\phi^{(3)} = |41|3$. Now regard the numbers between two consecutive bars as "blocks", and in each block, move the last element to the beginning, and finally remove all bars. We end up with $\phi^{(3)} = 143$.

Proceeding inductively, we begin by letting $\phi^{(i)}$ be the result of adding σ_i to the end of $\phi^{(i-1)}$. Then if if $\sigma_i > \sigma_{i-1}$, draw a bar after each element of $\phi^{(i)}$ which is greater than σ_i, while if $\sigma_i < \sigma_{i-1}$, draw a bar after each element of $\phi^{(i)}$ which is less than σ_i. Also draw a bar before $\phi_1^{(i)}$. Then in each block, move the last element to the beginning, and finally remove all bars. If $\sigma = 4137562$ the successive stages of the algorithm yield

$$\phi^{(3)} = 143$$

$$\phi^{(4)} = |1|4|3|7 \rightarrow 1437$$

$$\phi^{(5)} = |1437|5 \rightarrow 71435$$

$$\phi^{(6)} = |71|4|3|5|6 \rightarrow 174356$$

$$\phi^{(7)} = |17|4|3|5|6|2 \rightarrow 7143562$$

and so $\phi(4137562) = 7143562$.

PROPOSITION 1.1.1. *We have maj$(\sigma) = $ inv$(\phi(\sigma))$. Furthermore, Ides$(\sigma) = $ Ides$(\phi(\sigma))$, and ϕ fixes σ_n.*

PROOF. We claim inv$(\phi^{(k)}) = $ maj$(\sigma_1 \cdots \sigma_k)$ for $1 \leq k \leq n$. Clearly this is true for $k \leq 2$. Assume it is true for $k < j$, where $2 < j \leq n$. If $\sigma_j > \sigma_{j-1}$, maj$(\sigma_1 \cdots \sigma_j) = $ maj$(\sigma_1 \cdots \sigma_{j-1}) + j - 1$. One the other hand, for each block arising in the procedure creating $\phi^{(j)}$, the last element is greater than σ_j, which creates a new inversion, and when it is moved to the beginning of the block, it also creates a new inversion with each element in its block. It follows that inv$(\phi^{(j-1)}) = $ inv$(\phi^{(j)}) + j - 1$. Similar remarks hold if $\sigma_j < \sigma_{j-1}$. In this case maj$(\sigma_1 \cdots \sigma_{j-1}) = $ maj$(\sigma_1 \cdots \sigma_j)$. Also, each element of ϕ which is not the last element in its block is larger than σ_j, which creates a new inversion, but a corresponding inversion between this element and the last element in its block is lost when we cycle the last element to the beginning. Hence inv$(\phi^{(j-1)}) = $ inv$(\phi^{(j)})$ and the first part of the claim follows.

Note that Ides(σ) equals the set of all i, $1 \leq i < n$ such that $i+1$ occurs before i in σ. In order for the ϕ map to change this set, at some stage, say when creating $\phi^{(j)}$, we must move i from the end of a block to the beginning, passing $i-1$ or $i+1$ along the way. But this could only happen if σ_j is strictly between i and either $i-1$ or $i+1$, an impossibility. $\square$

Let $\beta = \phi^{-1}$, and begin by setting $\beta^{(1)} = \sigma$. Then if $\sigma_n > \sigma_1$, draw a bar *before* each number in $\beta^{(1)}$ which is less than σ_n, and also before σ_n. If $\sigma_n < \sigma_1$, draw a bar before each number in $\beta^{(1)}$ which is greater than σ_n, and also before σ_n. Next move each number at the beginning of a block to the end of the block.

The last letter of β is now fixed. Next set $\beta^{(2)} = \beta^{(1)}$, and compare the $n-1$st letter with the first, creating blocks as above, and draw an extra bar before the $n-1$st letter. For example, if $\sigma = 7143562$ the successive stages of the β algorithm yield

$$\beta^{(1)} = |71|4|3|5|6|2 \rightarrow 1743562$$

$$\beta^{(2)} = |17|4|3|5|62 \rightarrow 7143562$$

$$\beta^{(3)} = |7143|562 \rightarrow 1437562$$

$$\beta^{(4)} = |1|4|3|7562 \rightarrow 1437562$$

$$\beta^{(5)} = |14|37562 \rightarrow 4137562$$

$$\beta^{(6)} = \beta^{(7)} = 4137562$$

and so $\phi^{-1}(7143562) = 4137562$. Notice that at each stage we are reversing the steps of the ϕ algorithm, and it is easy to see this holds in general.

An *involution* on a set S is a bijective map from S to S whose square is the identity. Foata and Schützenberger [**FS78**] showed that the map $i\phi i\phi^{-1}i$, where i is the inverse map on permutations, is an involution on S_n which interchanges inv and maj.

For $n, k \in \mathbb{N}$, let

(1.4) $$\begin{bmatrix} n \\ k \end{bmatrix} = \frac{[n]!}{[k]![n-k]!} = \frac{(1-q^n)(1-q^{n-1})\cdots(1-q^{n-k+1})}{(1-q^k)(1-q^{k-1})\cdots(1-q)}$$

denote the Gaussian polynomial. These are special cases of more general objects known as q-binomial coefficients, which are defined for $x \in \mathbb{R}$ as

(1.5) $$\begin{bmatrix} x \\ k \end{bmatrix} = \frac{(q^{x-k+1}; q)_k}{(q; q)_k},$$

where $(a; q)_k = (a)_k = (1-a)(1-qa)\cdots(1-q^{k-1}a)$ is the "q-rising factorial".

A partition λ is a nonincreasing finite sequence $\lambda_1 \geq \lambda_2 \geq \ldots$ of positive integers. λ_i is called the ith part of λ. We let $\ell(\lambda)$ denote the number of parts, and $|\lambda| = \sum_i \lambda_i$ the sum of the parts. For various formulas it will be convenient to assume $\lambda_j = 0$ for $j > \ell(\lambda)$. The *Ferrers graph* of λ is an array of unit squares, called cells, with λ_i cells in the ith row, with the first cell in each row left-justified. We often use λ to refer to its Ferrers graph, We define the conjugate partition, λ' as the partition of whose Ferrers graph is obtained from λ by reflecting across the diagonal $x = y$. See Figure 1. for example $(i, j) \in \lambda$ refers to a cell with (column, row) coordinates (i, j), with the lower left-hand-cell of λ having coordinates $(1, 1)$. The notation $x \in \lambda$ means x is a cell in λ. For technical reasons we say that 0 has one partition, the emptyset $\emptyset$, with $\ell(\emptyset) = 0 = |\emptyset|$.

FIGURE 1. On the left, the Ferrers graph of the partition $(4, 3, 2, 2)$, and on the right, that of its conjugate $(4, 3, 2, 2)' = (4, 4, 2, 1)$.

The following result shows the Gaussian polynomials are in fact polynomials in q, which is not obvious from their definition.

THEOREM 1.2. *For $n, k \in \mathbb{N}$,*

(1.6) $$\begin{bmatrix} n + k \\ k \end{bmatrix} = \sum_{\lambda \subseteq n^k} q^{|\lambda|},$$

where the sum is over all partitions λ whose Ferrers graph fits inside a $k \times n$ rectangle, i.e. for which $\lambda_1 \leq n$ and $\ell(\lambda) \leq k$.

PROOF. Let

$$P(n, k) = \sum_{\lambda \subseteq n^k} q^{|\lambda|}.$$

Clearly

$$(1.7) \qquad P(n,k) = \sum_{\substack{\lambda \subseteq n^k \\ \lambda_1 = n}} q^{|\lambda|} + \sum_{\substack{\lambda \subseteq n^k \\ \lambda_1 \leq n-1}} q^{|\lambda|} = q^n P(n, k-1) + P(n-1, k).$$

On the other hand

$$
\begin{aligned}
q^n \begin{bmatrix} n+k-1 \\ k-1 \end{bmatrix} + \begin{bmatrix} n-1+k \\ k \end{bmatrix} &= q^n \frac{[n+k-1]!}{[k-1]![n]!} + \frac{[n-1+k]!}{[k]![n-1]!} \\
&= \frac{q^n[k][n+k-1]! + [n-1+k]![n]}{[k]![n]!} \\
&= \frac{[n+k-1]!}{[k]![n]!} (q^n (1 + q + \ldots + q^{k-1}) + 1 + q + \ldots + q^{n-1}) \\
&= \frac{[n+k]!}{[k]![n]!}.
\end{aligned}
$$

Since $P(n,0) = P(0,k) = 1$, both sides of (1.6) thus satisfy the same recurrence and initial conditions. □

Given $\alpha = (\alpha_0, \ldots, \alpha_s) \in \mathbb{N}^{s+1}$, let

$$\{0^{\alpha_0} 1^{\alpha_1} \cdots s^{\alpha_s}\}$$

denote the multiset with α_i copies of i, where $\alpha_0 + \ldots + \alpha_s = n$. We let M_α denote the set of all permutations of this multiset and refer to α as the *weight* of any given one of these words. We extend the definitions of inv and maj as follows

$$(1.8) \qquad \mathrm{inv}(\sigma) = \sum_{\substack{i < j \\ \sigma_i > \sigma_j}} 1, \quad \mathrm{maj}(\sigma) = \sum_{\substack{i \\ \sigma_i > \sigma_{i+1}}} i.$$

Also let

$$(1.9) \qquad \begin{bmatrix} n \\ \alpha_0, \ldots, \alpha_s \end{bmatrix} = \frac{[n]!}{[\alpha_0]! \cdots [\alpha_s]!}$$

denote the "q-multinomial coefficient".

The following result is due to MacMahon [**Mac60**].

THEOREM 1.3. *Both inv and maj are multiset Mahonian, i.e. given $\alpha \in \mathbb{N}^{s+1}$,*

$$(1.10) \qquad \sum_{\sigma \in M_\alpha} q^{inv(\sigma)} = \begin{bmatrix} n \\ \alpha_0, \ldots, \alpha_s \end{bmatrix} = \sum_{\sigma \in M_\alpha} q^{maj(\sigma)},$$

where the sums are over all elements $\sigma \in M_\alpha$.

REMARK 1.4. Foata's map also proves Theorem 1.3 bijectively. To see why, given a multiset permutation σ of $M(\beta)$ let σ' denote the *standardization* of σ, defined to be the permutation obtained by replacing the β_1's by the numbers 1 through β_1, in increasing order as we move left to right in σ, then replacing the β_2's by the numbers $\beta_1 + 1$ through $\beta_1 + \beta_2$, again in increasing order as we move left to right in σ, etc. For example, the standardization of 31344221 is 51678342. Note that

$$(1.11) \qquad \mathrm{Ides}(\sigma') \subseteq \{\beta_1, \beta_1 + \beta_2, \ldots\}$$

and in fact standardization gives a bijection between elements of $M(\beta)$ and permutations satisfying (1.11).

EXERCISE 1.5. If σ is a word of length n define the co-major index of σ as follows.

$$(1.12) \qquad \mathrm{comaj}(\sigma) = \sum_{\substack{\sigma_i > \sigma_{i+1} \\ 1 \le i < n}} n - i.$$

Show that Foata's map ϕ implies there is a bijective map $\mathrm{co}\phi$ on words of fixed weight such that

$$(1.13) \qquad \mathrm{comaj}(\sigma) = \mathrm{inv}(\mathrm{co}\phi(\sigma)).$$

The Catalan Numbers and Dyck Paths

A lattice path is a sequence of North $N(0,1)$ and East $E(1,0)$ steps in the first quadrant of the xy-plane, starting at the origin $(0,0)$ and ending at say (n,m). We let $L_{n,m}$ denote the set of all such paths, and $L_{n,m}^+$ the subset of $L_{n,m}$ consisting of paths which never go below the line $y = \frac{m}{n}x$. A Dyck path is an element of $L_{n,n}^+$ for some n.

Let $C_n = \frac{1}{n+1}\binom{2n}{n}$ denote the nth Catalan number, so

$$C_0, C_1, \ldots = 1, 1, 2, 5, 14, 42, \ldots.$$

See [**Sta99**, Ex. 6.19, p. 219] for a list of over 66 objects counted by the Catalan numbers. One of these is the number of elements of $L_{n,n}^+$. For $1 \le k \le n$, the number of Dyck paths from $(0,0)$ to (k,k) which touch the line $y = x$ only at $(0,0)$ and (k,k) is C_{k-1}, since such a path must begin with a N step, end with an E step, and never go below the line $y = x + 1$ as it goes from $(0,1)$ to $(k-1,k)$. The number of ways to extend such a path to (n,n) and still remain a Dyck path is clearly C_{n-k}. It follows that

$$(1.14) \qquad C_n = \sum_{k=1}^{n} C_{k-1} C_{n-k} \quad n \ge 1.$$

There are two natural q-analogues of C_n. Given $\pi \in L_{n,m}$, let $\sigma(\pi)$ be the element of $M_{(m,n)}$ resulting from the following algorithm. First initialize σ to the empty string. Next start at $(0,0)$, move along π and add a 0 to the end of $\sigma(\pi)$ every time a N step is encountered, and add a 1 to the end of $\sigma(\pi)$ every time an E step is encountered. Similarly, given $\sigma \in M_{(m,n)}$, let $\pi(\sigma)$ be the element of $L_{n,m}$ obtained by inverting the above algorithm. We call the transformation of π to σ or its inverse the *coding* of π or σ. Let $a_i(\pi)$ denote the number of complete squares, in the ith row from the bottom of π, which are to the right of π and to the left of the line $y = x$. We refer to $a_i(\pi)$ as the *length* of the ith row of π. Furthermore call $(a_1(\pi), a_2(\pi), \ldots, a_n(\pi))$ the *area vector* of π, and set $\mathrm{area}(\pi) = \sum_i a_i(\pi)$. For example, the path in Figure 2 has area vector $(0,1,1,2,1,2,0)$, and $\sigma(\pi) = 00100110011101$. By convention we say $L_{0,0}^+$ contains one path, the empty path $\emptyset$, with $\mathrm{area}(\emptyset) = 0$.

Let $M_{(m,n)}^+$ denote the elements σ of $M_{(m,n)}$ corresponding to paths in $L_{n,m}^+$. Paths in $M_{n,n}^+$ are thus characterized by the property that in any initial segment there are at least as many 0's as 1's. The first q-analogue of C_n is given by the following.

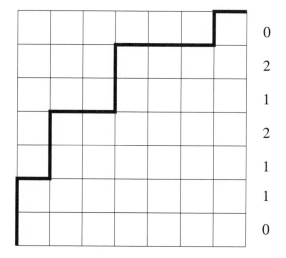

0
2
1
2
1
1
0

FIGURE 2. A Dyck path, with row lengths on the right. The area statistic is $1 + 1 + 2 + 1 + 2 = 7$.

THEOREM 1.6. *(MacMahon [**Mac60**, p. 214])*

$$(1.15) \qquad \sum_{\pi \in L_{n,n}^+} q^{maj(\sigma(\pi))} = \frac{1}{[n+1]} \begin{bmatrix} 2n \\ n \end{bmatrix}.$$

PROOF. We give a *bijective* proof, taken from [**FH85**]. Let $M_{(m,n)}^- = M_{(m,n)} \setminus M_{(m,n)}^+$, and let $L_{n,m}^- = L_{n,m} \setminus L_{n,m}^+$ be the corresponding set of lattice paths. Given a path $\pi \in L_{n,n}^-$, let P be the point with smallest x-coordinate among those lattice points (i,j) in π for which $i-j$ is maximal, i.e. whose distance from the line $y = x$ in a SE direction is maximal. (Since $\pi \in L_{n,n}^-$, this maximal value of $i-j$ is positive.) Let P' be the lattice point on π before P. There must be a E step connecting P' to P. Change this into a N step and shift the remainder of the path after P up one unit and left one unit. We now have a path $\phi(\pi)$ from $(0,0)$ to $(n-1, n+1)$, and moreover $maj(\sigma(\phi(\pi))) = maj(\sigma(\pi)) - 1$.

It is easy to see that this map is invertible. Given a lattice path π' from $(0,0)$ to $(n-1, n+1)$, let P' be the point with maximal x-coordinate among those lattice points (i,j) in π' for which $j-i$ is maximal. Thus

$$(1.16) \qquad \sum_{\sigma \in M_{(n,n)}^-} q^{\mathrm{maj}(\sigma)} = \sum_{\sigma' \in M_{(n+1,n-1)}} q^{\mathrm{maj}(\sigma')+1} = q \begin{bmatrix} 2n \\ n+1 \end{bmatrix},$$

using (1.10). Hence

$$(1.17) \qquad \sum_{\pi \in L_{n,n}^+} q^{\mathrm{maj}(\sigma(\pi))} = \sum_{\sigma \in M_{(n,n)}} q^{\mathrm{maj}(\sigma)} - \sum_{\sigma \in M_{(n,n)}^-} q^{\mathrm{maj}(\sigma)}$$

$$(1.18) \qquad\qquad = \begin{bmatrix} 2n \\ n \end{bmatrix} - q \begin{bmatrix} 2n \\ n+1 \end{bmatrix} = \frac{1}{[n+1]} \begin{bmatrix} 2n \\ n \end{bmatrix}.$$

$\square$

The second natural q-analogue of C_n was studied by Carlitz and Riordan [**CR64**]. They define

$$(1.19) \qquad C_n(q) = \sum_{\pi \in L_{n,n}^+} q^{\text{area}(\pi)}.$$

PROPOSITION 1.6.1.

$$(1.20) \qquad C_n(q) = \sum_{k=1}^{n} q^{k-1} C_k(q) C_{n-k}(q), \quad n \geq 1.$$

PROOF. As in the proof of (1.14), we break up our path π according to the "point of first return" to the line $y = x$. If this occurs at (k, k), then the area of the part of π from $(0, 1)$ to $(k-1, k)$, when viewed as an element of $L_{k-1,k-1}^+$, is $k - 1$ less than the area of this portion of π when viewed as a path in $L_{n,n}^+$. □

EXERCISE 1.7. Define a *co-inversion* of σ to be a pair (i, j) with $i < j$ and $\sigma_i < \sigma_j$. Show

$$(1.21) \qquad C_n(q) = \sum_{\pi \in L_{n,n}^+} q^{\text{coinv}(\sigma(\pi)) - \binom{n+1}{2}},$$

where $\text{coinv}(\sigma)$ is the number of co-inversions of σ.

The q-Vandermonde Convolution

Let

$$(1.22) \qquad {}_{p+1}\phi_p \left(\begin{matrix} a_1, & a_2, & \ldots, & a_{p+1} \\ & b_1, & \ldots, & b_p \end{matrix} ; q; z \right) = \sum_{n=0}^{\infty} \frac{(a_1)_n \cdots (a_{p+1})_n}{(q)_n (b_1)_n \cdots (b_p)_n} z^n$$

denote the basic hypergeometric series. A good general reference for this subject is [**GR04**]. The following result is known as Cauchy's q-binomial series.

THEOREM 1.8.

$$(1.23) \qquad {}_1\phi_0 \left(\begin{matrix} a \\ - \end{matrix} ; q; z \right) = \sum_{n=0}^{\infty} \frac{(a)_n}{(q)_n} z^n = \frac{(az)_\infty}{(z)_\infty}, \quad |z| < 1, |q| < 1,$$

where $(a; q)_\infty = (a)_\infty = \prod_{i=0}^{\infty} (1 - aq^i)$.

PROOF. The following proof is based on the proof in [**GR04**, Chap. 1]. Let

$$F(a, z) = \sum_{n=0}^{\infty} \frac{(a)_n}{(q)_n} z^n.$$

Then

$$(1.24) \qquad F(a, z) - F(a, qz) = (1 - a) z F(aq, z)$$

and

$$(1.25) \qquad F(a, z) - F(aq, z) = -az F(aq, z).$$

Eliminating $F(aq, z)$ from (1.24) and (1.25) we get

$$F(a, z) = \frac{(1 - az)}{(1 - z)} F(a, qz).$$

Iterating this n times then taking the limit as $n \to \infty$ we get

$$F(a, z) = \lim_{n \to \infty} \frac{(az; q)_n}{(z; q)_n} F(a, q^n z)$$

(1.26)
$$= \frac{(az; q)_\infty}{(z; q)_\infty} F(a, 0) = \frac{(az; q)_\infty}{(z; q)_\infty}.$$

$\square$

COROLLARY 1.8.1. *The "q-binomial theorem".*

(1.27)
$$\sum_{k=0}^{n} q^{\binom{k}{2}} \begin{bmatrix} n \\ k \end{bmatrix} z^k = (-z; q)_n$$

and

(1.28)
$$\sum_{k=0}^{\infty} \begin{bmatrix} n + k \\ k \end{bmatrix} z^k = \frac{1}{(z; q)_{n+1}}.$$

PROOF. To prove (1.27), set $a = q^{-n}$ and $z = -zq^n$ in (1.23) and simplify. To prove (1.28), let $a = q^{n+1}$ in (1.23) and simplify. $\square$

For any function $f(z)$, let $f(z)|_{z^k}$ denote the coefficient of z^k in the Maclaurin series for $f(z)$.

COROLLARY 1.8.2.

(1.29)
$$\sum_{k=0}^{h} q^{(n-k)(h-k)} \begin{bmatrix} n \\ k \end{bmatrix} \begin{bmatrix} m \\ h - k \end{bmatrix} = \begin{bmatrix} m + n \\ h \end{bmatrix}.$$

PROOF. By (1.27),

$$q^{\binom{h}{2}} \begin{bmatrix} m + n \\ h \end{bmatrix} = \prod_{k=0}^{m+n-1} (1 + zq^k)|_{z^h}$$

$$= \prod_{k=0}^{n-1} (1 + zq^k) \prod_{j=0}^{m-1} (1 + zq^n q^j)|_{z^h}$$

$$= (\sum_{k=0}^{n-1} q^{\binom{k}{2}} \begin{bmatrix} n \\ k \end{bmatrix} z^k)(\sum_{j=0}^{m-1} q^{\binom{j}{2}} \begin{bmatrix} m \\ j \end{bmatrix} (zq^n)^j)|_{z^h}$$

$$= \sum_{k=0}^{h} q^{\binom{k}{2}} \begin{bmatrix} n \\ k \end{bmatrix} q^{\binom{h-k}{2}} \begin{bmatrix} m \\ h - k \end{bmatrix} (q^n)^{h-k}.$$

The result now reduces to the identity

$$\binom{k}{2} + \binom{h-k}{2} + n(h-k) - \binom{h}{2} = (n-k)(h-k).$$

$\square$

COROLLARY 1.8.3.

(1.30)
$$\sum_{k=0}^{h} q^{(m+1)k} \begin{bmatrix} n - 1 + k \\ k \end{bmatrix} \begin{bmatrix} m + h - k \\ h - k \end{bmatrix} = \begin{bmatrix} m + n + h \\ h \end{bmatrix}.$$

PROOF. By (1.28),

$$
\begin{aligned}
\begin{bmatrix} m+n+h \\ h \end{bmatrix} &= \frac{1}{(z)_{m+n+1}}\Big|_{z^h} \\
&= \frac{1}{(z)_{m+1}}\frac{1}{(zq^{m+1})_n}\Big|_{z^h} \\
&= \left(\sum_{j=0}^{h} z^j \begin{bmatrix} m+j \\ j \end{bmatrix}\right)\left(\sum_{k=0}^{h}(zq^{m+1})^k \begin{bmatrix} n-1+k \\ k \end{bmatrix}\right)\Big|_{z^h} \\
&= \sum_{k=0}^{h} q^{(m+1)k} \begin{bmatrix} n-1+k \\ k \end{bmatrix}\begin{bmatrix} m+h-k \\ h-k \end{bmatrix}.
\end{aligned}
$$

$\square$

Both (1.29) and (1.30) are special cases of the following result, known as the q-Vandermonde convolution. For a proof see [**GR04**, Chap. 1].

THEOREM 1.9. *Let $n \in \mathbb{N}$. Then*

$$
(1.31) \qquad {}_2\phi_1\left(\begin{matrix} a, & q^{-n} \\ & c \end{matrix}; q; q\right) = \frac{(c/a)_n}{(c)_n}a^n.
$$

EXERCISE 1.10. By reversing summation in (1.31), show that

$$
(1.32) \qquad {}_2\phi_1\left(\begin{matrix} a, & q^{-n} \\ & c \end{matrix}; q; cq^n/a\right) = \frac{(c/a)_n}{(c)_n}.
$$

EXERCISE 1.11. Show Newton's binomial series

$$
(1.33) \qquad \sum_{n=0}^{\infty} \frac{a(a+1)\cdots(a+n-1)}{n!}z^n = \frac{1}{(1-z)^a}, \quad |z| < 1,\ a \in \mathbb{R}
$$

can be derived from (1.23) by replacing a by q^a and letting $q \to 1^-$. For simplicity you can assume $a, z \in \mathbb{R}$.

Symmetric Functions

The Basics. Given $f(x_1, \ldots, x_n) \in K[x_1, x_2, \ldots, x_n]$ for some field K, and $\sigma \in S_n$, let

$$
(1.34) \qquad \sigma f = f(x_{\sigma_1}, \ldots, x_{\sigma_n}).
$$

We say f is a *symmetric function* if $\sigma f = f$, $\forall \sigma \in S_n$. It will be convenient to work with more general functions f depending on countably many indeterminates $x_1, x_2, \ldots$, indicated by $f(x_1, x_2, \ldots)$, in which case we view f as a formal power series in the x_i, and say it is a symmetric function if it is invariant under any permutation of the variables. The standard references on this topic are [**Sta99**, Chap. 7] and [**Mac95**]. We will often let X_n and X stand for the set of variables $\{x_1, \ldots, x_n\}$ and $\{x_1, x_2, \ldots\}$, respectively.

We let Λ denote the ring of symmetric functions in $x_1, x_2, \ldots$ and Λ^n the sub vector space consisting of symmetric functions of homogeneous degree n. The most basic symmetric functions are the monomial symmetric functions, which depend on a partition λ in addition to a set of variables. They are denoted by $m_\lambda(X) = m_\lambda(x_1, x_2, \ldots)$. In a symmetric function it is typical to leave off explicit mention of the variables, with a set of variables being understood from context, so $m_\lambda =$

$m_\lambda(X)$. We illustrate these first by means of examples. We let $\mathrm{Par}(n)$ denote the set of partitions of n, and use the notation $\lambda \vdash n$ as an abbreviation for $\lambda \in \mathrm{Par}(n)$.

EXAMPLE 1.12.

$$m_{1,1} = \sum_{i<j} x_i x_j$$
$$m_{2,1,1}(X_3) = x_1^2 x_2 x_3 + x_1 x_2^2 x_3 + x_1 x_2 x_3^2$$
$$m_2(X) = \sum_i x_i^2.$$

In general, $m_\lambda(X)$ is the sum of all distinct monomials in the x_i whose multiset of exponents equals the multiset of parts of λ. Any element of Λ can be expressed uniquely as a linear combination of the m_λ.

We let 1^n stand for the partition consisting of n parts of size 1. The function m_{1^n} is called the nth *elementary symmetric function*, which we denote by e_n. Then

$$(1.35) \qquad \prod_{i=1}^{\infty}(1 + z x_i) = \sum_{n=0}^{\infty} z^n e_n, \quad e_0 = 1.$$

Another important special case is $m_n = \sum_i x_i^n$, known as the *power-sum symmetric functions*, denoted by p_n. We also define the *complete homogeneous symmetric functions* h_n, by $h_n = \sum_{\lambda \vdash n} m_\lambda$, or equivalently

$$(1.36) \qquad \frac{1}{\prod_{i=1}^{\infty}(1 - z x_i)} = \sum_{n=0}^{\infty} z^n h_n, \quad h_0 = 1.$$

For $\lambda \vdash n$, we define $e_\lambda = \prod_i e_{\lambda_i}$, $p_\lambda = \prod_i p_{\lambda_i}$, and $h_\lambda = \prod_i h_{\lambda_i}$. For example,

$$e_{2,1} = \sum_{i<j} x_i x_j \sum_k x_k = m_{2,1} + 3 m_{1,1,1}$$
$$p_{2,1} = \sum_i x_i^2 \sum_j x_j = m_3 + m_{2,1}$$
$$h_{2,1} = \left(\sum_i x_i^2 + \sum_{i<j} x_i x_j\right) \sum_k x_k = m_3 + 2 m_{2,1} + 3 m_{1,1,1}.$$

It is known that $\{e_\lambda, \lambda \vdash n\}$ forms a basis for Λ^n, and so do $\{p_\lambda, \lambda \vdash n\}$ and $\{h_\lambda, \lambda \vdash n\}$.

DEFINITION 1.13. Two simple functions on partitions we will often use are

$$n(\lambda) = \sum_i (i-1)\lambda_i = \sum_i \binom{\lambda_i'}{2}$$
$$z_\lambda = \prod_i i^{n_i} n_i!,$$

where $n_i = n_i(\lambda)$ is the number of parts of λ equal to i.

EXERCISE 1.14. Use (1.35) and (1.36) to show that

$$e_n = \sum_{\lambda \vdash n} \frac{(-1)^{n-\ell(\lambda)} p_\lambda}{z_\lambda},$$
$$h_n = \sum_{\lambda \vdash n} \frac{p_\lambda}{z_\lambda}.$$

We let ω denote the standard endomorphism $\omega : \Lambda \to \Lambda$ defined by

$$(1.37) \qquad \omega(p_k) = (-1)^{k-1} p_k.$$

Thus ω is an involution with $\omega(p_\lambda) = (-1)^{|\lambda|-\ell(\lambda)} p_\lambda$, and by Exercise 1.14, $\omega(e_n) = h_n$, and more generally $\omega(e_\lambda) = h_\lambda$.

For $f \in \Lambda$, the special value $f(1, q, q^2, \ldots, q^{n-1})$ is known as the principal specialization (of order n) of f.

THEOREM 1.15.

$$(1.38) \qquad e_m(1, q, \ldots, q^{n-1}) = q^{\binom{m}{2}} \begin{bmatrix} n \\ m \end{bmatrix}$$

$$(1.39) \qquad h_m(1, q, \ldots, q^{n-1}) = \begin{bmatrix} n - 1 + m \\ m \end{bmatrix}$$

$$(1.40) \qquad p_m(1, q, \ldots, q^{n-1}) = \frac{1 - q^{nm}}{1 - q^m}.$$

PROOF. The principal specializations for e_m and h_m follow directly from (1.27), (1.28), (1.35) and (1.36). $\qquad \square$

REMARK 1.16. The principal specialization of m_λ doesn't have a particularly simple description, although if ps_n^1 denotes the set of n variables, each equal to 1, then [**Sta99**, p. 303]

$$(1.41) \qquad m_\lambda(ps_n^1) = \binom{n}{n_1, n_2, n_3, \ldots},$$

where again n_i is the multiplicity of the number i in the multiset of parts of λ.

REMARK 1.17. Identities like

$$h_{2,1} = m_3 + m_{2,1} + m_{1,1,1}$$

appear at first to depend on a set of variables, but it is customary to view them as polynomial identities in the p_λ. Since the p_k are algebraically independent, we can specialize them to whatever we please, forgetting about the original set of variables X.

EXERCISE 1.18.   (1) Show that

$$(1.42) \qquad \prod_{i=1}^{\infty} (1 + x_i t_1 + x_i^2 t_2 + x_i^3 t_3 + \ldots) = \sum_{\lambda \in \mathrm{Par}} m_\lambda(X) \prod_i t_{\lambda_i}.$$

Use this to prove (1.41).

(2) If $f \in \Lambda$ and $x \in \mathbb{R}$, let $f(ps_x^1)$ denote the polynomial in x obtained by first expanding f in terms of the p_λ, then replacing each p_k by x. Use Theorem 1.15 and 1.41 to show that

$$e_m(ps_x^1) = \binom{x}{m}$$

$$h_m(ps_x^1) = \binom{x + m - 1}{m}$$

$$m_\lambda(ps_x^1) = \binom{x}{n_1, n_2, n_3, \ldots}.$$

We define the Hall scalar product, a bilinear form from $\Lambda \times \Lambda \to \mathbb{Q}$, by

(1.43) $$\langle p_\lambda, p_\beta \rangle = z_\lambda \chi(\lambda = \beta),$$

where for any logical statement L

(1.44) $$\chi(L) = \begin{cases} 1 & \text{if } L \text{ is true} \\ 0 & \text{if } L \text{ is false.} \end{cases}$$

Clearly $< f, g >=< g, f >$. Also, $< \omega f, \omega g >=< f, g >$, which follows from the definition if $f = p_\lambda, g = p_\beta$, and by bilinearity for general f, g since the p_λ form a basis for Λ.

THEOREM 1.19. *The h_λ and the m_β are dual with respect to the Hall scalar product, i.e.*

(1.45) $$\langle h_\lambda, m_\beta \rangle = \chi(\lambda = \beta).$$

PROOF. See [**Mac95**] or [**Sta99**]. □

For any $f \in \Lambda$, and any basis $\{b_\lambda, \lambda \in \text{Par}\}$ of Λ, let $f|_{b_\lambda}$ denote the coefficient of b_λ when f is expressed in terms of the b_λ. Then (1.45) implies

COROLLARY 1.19.1.

(1.46) $$\langle f, h_\lambda \rangle = f|_{m_\lambda}.$$

Tableaux and Schur Functions. Given $\lambda, \mu \in \text{Par}(n)$, a semi-standard Young tableaux (or SSYT) of shape λ and weight μ is a filling of the cells of the Ferrers graph of λ with the elements of the multiset $\{1^{\mu_1} 2^{\mu_2} \cdots\}$, so that the numbers weakly increase across rows and strictly increase up columns. Let $SSYT(\lambda, \mu)$ denote the set of these fillings, and $K_{\lambda,\mu}$ the cardinality of this set. The $K_{\lambda,\mu}$ are known as the Kostka numbers. Our definition also makes sense if our weight is a weak composition of n, i.e. any finite sequence of nonnegative integers whose sum is n. For example, $K_{(3,2),(2,2,1)} = K_{(3,2),(2,1,2)} = K_{(3,2),(1,2,2)} = 2$ as in Figure 3.

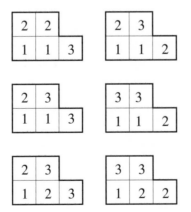

FIGURE 3. Some SSYT of shape $(3, 2)$.

If the Ferrers graph of a partition β is contained in the Ferrers graph of λ, denoted $\beta \subseteq \lambda$, let λ/β refer to the subset of cells of λ which are not in β. This is referred to as a *skew shape*. Define a SSYT of shape λ/β and weight ν, where

$|\nu| = |\lambda| - |\beta|$, to be a filling of the cells of λ/β with elements of $\{1^{\nu_1}2^{\nu_2}\cdots\}$, again with weak increase across rows and strict increase up columns. The number of such tableaux is denoted $K_{\lambda/\beta,\nu}$.

Let wcomp(μ) denote the set of all weak compositions whose multiset of nonzero parts equals the multiset of parts of μ. It follows easily from Figure 3 that $K_{(3,2),\alpha} = 2$ for all $\alpha \in$ wcomp$(2,2,1)$. Hence

$$(1.47) \qquad \sum_{\alpha,T}\prod_i x_i^{\alpha_i} = 2m_{(2,2,1)},$$

where the sum is over all tableaux T of shape $(3,2)$ and weight some element of wcomp$(2,2,1)$.

This is a special case of a more general phenomenon. For $\lambda \in$ Par(n), define

$$(1.48) \qquad s_\lambda = \sum_{\alpha,T}\prod_i x_i^{\alpha_i},$$

where the sum is over all weak compositions α of n, and all possible tableaux T of shape λ and weight α. Then

$$(1.49) \qquad s_\lambda = \sum_{\mu \vdash n} K_{\lambda,\mu} m_\mu.$$

The s_λ are called *Schur functions*, and are fundamental to the theory of symmetric functions. Two special cases of (1.49) are $s_n = h_n$ (since $K_{n,\mu} = 1$ for all $\mu \in$ Par(n)) and $s_{1^n} = e_n$ (since $K_{1^n,\mu} = \chi(\mu = 1^n)$).

A SSYT of weight 1^n is called *standard*, or a SYT. The set of SYT of shape λ is denoted $SYT(\lambda)$. For $(i,j) \in \lambda$, let the "content" of (i,j), denoted $c(i,j)$, be $j - i$. Also, let $h(i,j)$ denote the "hook length" of (i,j), defined as the number of cells to the right of (i,j) in row i plus the number of cells above (i,j) in column j plus 1. For example, if $\lambda = (5,5,3,3,1)$, $h(2,2) = 6$. It is customary to let f^λ denote the number of SYT of shape λ, i.e. $f^\lambda = K_{\lambda,1^n}$. There is a beautiful formula for f^λ, namely

$$(1.50) \qquad f^\lambda = \frac{n!}{\prod_{(i,j)\in\lambda} h(i,j)}.$$

Below we list some of the important properties of Schur functions.

THEOREM 1.20. *Let* $\lambda, \mu \in$ *Par. Then*

(1) *The Schur functions are orthonormal with respect to the Hall scalar product, i.e.*

$$< s_\lambda, s_\mu > = \chi(\lambda = \mu).$$

Thus for any $f \in \Lambda$,

$$< f, s_\lambda > = f|_{s_\lambda}.$$

(2) *Action by* ω:

$$\omega(s_\lambda) = s_{\lambda'}.$$

(3) *(The Jacobi-Trudi identity)*

$$s_\lambda = det(h_{\lambda_i - i + j})_{i,j=1}^{\ell(\lambda)},$$

where we set $h_0 = 1$ and $h_k = 0$ for $k < 0$. For example,

$$s_{2,2,1} = \begin{vmatrix} h_2 & h_3 & h_4 \\ h_1 & h_2 & h_3 \\ 0 & 1 & h_1 \end{vmatrix}$$

$$= h_{2,2,1} - h_{3,2} - h_{3,1,1} + h_{4,1}.$$

(4) *(The Pieri rule). Let $k \in \mathbb{N}$. Then*

$$s_\lambda h_k = \sum_\gamma s_\gamma,$$

where the sum is over all γ whose Ferrers graph contains λ with $|\gamma/\lambda| = k$ and such that γ/λ is a "horizontal strip", i.e. has no two cells in any column. Also,

$$s_\lambda e_k = \sum_\gamma s_\gamma,$$

where the sum is over all partitions γ whose Ferrers graph contains λ with $|\gamma/\lambda| = k$ and such that γ/λ is a "vertical strip", i.e. has no two cells in any row. For example,

$$s_{2,1} h_2 = s_{4,1} + s_{3,2} + s_{3,1,1} + s_{2,2,1}$$

$$s_{3,1} e_2 = s_{4,2} + s_{4,1,1} + s_{3,2,1} + s_{3,1,1,1}.$$

(5) *(Principal Specialization).*

$$s_\lambda(1, q, q^2, \ldots, q^{n-1}) = q^{n(\lambda)} \prod_{(i,j) \in \lambda} \frac{[n + c(i,j)]}{[h(i,j)]},$$

and taking the limit as $n \to \infty$ we get

$$s_\lambda(1, q, q^2, \ldots) = q^{n(\lambda)} \prod_{(i,j) \in \lambda} \frac{1}{1 - q^{h(i,j)}}.$$

(6) *For any two alphabets Z, W,*

(1.51) $$s_\lambda(Z + W) = \sum_{\beta \subseteq \lambda} s_\beta(Z) s_{\lambda'/\beta'}(W).$$

(7) *(The Littlewood-Richardson rule) For all partitions λ, β,*

$$s_\lambda s_\beta = \sum_\gamma s_\gamma c^\gamma_{\lambda,\beta},$$

*where $c^\gamma_{\lambda,\beta}$ is the number of SSYT T of skew shape γ/β and weight λ such that if we form the word $\sigma_1 \sigma_2 \cdots$ by reading the entries of T across rows from right to left, and from bottom to top, then in any initial segment $\sigma_1 \cdots \sigma_j$ there are at least as many i's as $i+1$'s, for each $1 \le i$. (Such words are known as lattice permutations, and the corresponding SSYT are called Yamanouchi). Note that this rule contains the Pieri rules above as special cases. For a proof of the Littlewood-Richardson rule see [**Mac88**, Chap. 1].*

EXERCISE 1.21. Let $0 \leq 2k \leq n$. Show

$$\sum_{\lambda} f^{\lambda} = \binom{n}{k},$$

where the sum is over all $\lambda \in \mathrm{Par}(n)$ with $\ell(\lambda) \leq 2$ and $\lambda_1 \geq n - k$.

Statistics on Tableaux. There is a q-analogue of the Kostka numbers, denoted by $K_{\lambda,\mu}(q)$, which has many applications in representation theory and the combinatorics of tableaux. Originally defined algebraically in an indirect fashion, the $K_{\lambda,\mu}(q)$ are polynomials in q which satisfy $K_{\lambda,\mu}(1) = K_{\lambda,\mu}$. Foulkes [**Fou74**] conjectured that there should be a statistic $\mathrm{stat}(T)$ on SSYT T of shape λ and weight μ such that

(1.52) $$K_{\lambda,\mu}(q) = \sum_{T \in SSYT(\lambda)} q^{\mathrm{stat}(T)}.$$

This conjecture was resolved by Lascoux and Schützenberger [**LS78**], who found a statistic *charge* to generate these polynomials. Butler [**But94**] provided a detailed account of their proof, filling in a lot of missing details. A short proof, based on the new combinatorial formula for Macdonald polynomials, is contained in Appendix A.

Assume we have a tableau $T \in SSYT(\lambda, \mu)$ where $\mu \in \mathrm{Par}$. It will be more convenient for us to describe a slight modification of charge(T), called cocharge(T), which is defined as $n(\mu) -$ charge. The *reading word* read(T) of T is obtained by reading the entries in T from left to right in the top row of T, then continuing left to right in the second row from the top of T, etc. For example, the tableau in the upper-left of Figure 3 has reading word 22113. To calculate cocharge(T), perform the following algorithm on read(T).

ALGORITHM 1.22.   (1) *Start at the end of read(T) and scan left until you encounter a 1 - say this occurs at spot i_1, so $read(T)_{i_1} = 1$. Then start there and scan left until you encounter a 2. If you hit the end of read(T) before finding a 2, loop around and continue searching left, starting at the end of read(T). Say the first 2 you find equals $read(T)_{i_2}$. Now iterate, start at i_2 and search left until you find a 3, etc. Continue in this way until you have found $4, 5, \ldots, \mu_1$, with μ_1 occurring at spot i_{μ_1}. Then the first subword of $textread(T)$ is defined to be the elements of the set $\{read(T)_{i_1}, \ldots, read(T)_{i_{\mu_1}}\}$, listed in the order in which they occur in read(T) if we start at the beginning of read(T) and move left to right. For example, if $read(T) = 21613244153$ then the first subword equals 632415, corresponding to places $3, 5, 6, 8, 9, 10$ of read(T).*

Next remove the elements of the first subword from read(T) and find the first subword of what's left. Call this the second subword. Remove this and find the first subword in what's left and call this the third subword of read(T), etc. For the word 21613244153, the subwords are 632415, 2143, 1.

(2) *The value of charge(T) will be the sum of the values of charge on each of the subwords of $rw(T)$. Thus it suffices to assume $rw(T) \in S_m$ for some m, in which case we set*

$$cocharge(rw(T)) = comaj(rw(T)^{-1}),$$

where $\text{read}(T)^{-1}$ *is the usual inverse in* S_m, *with comaj as in (1.12).*
(Another way of describing cocharge(read(T)) is the sum of $m - i$ *over all*
i *for which* $i + 1$ *occurs before* i *in read(T).) For example, if* $\sigma = 632415$,
then $\sigma^{-1} = 532461$ *and* $\text{cocharge}(\sigma) = 5 + 4 + 1 = 10$, *and finally*

$$\text{cocharge}(21613244153) = 10 + 4 + 0 = 14.$$

Note that to compute charge, we could create subwords in the same manner, and count $m - i$ for each i with $i + 1$ occurring to the right of i instead of to the left. For $\lambda, \mu \in \text{Par}(n)$ we set

$$(1.53) \qquad \tilde{K}_{\lambda,\mu}(q) = q^{n(\lambda)} K_{\lambda,\mu}(1/q)$$

$$= \sum_{T \in SSYT(\lambda,\mu)} q^{\text{cocharge}(T)}.$$

These polynomials have an interpretation in terms of representation theory which we describe in Chapter 2.

In addition to the cocharge statistic, there is a major index statistic on SYT which is often useful. Given a SYT tableau T of shape λ , define a descent of T to be a value of i, $1 \le i < |\lambda|$, for which $i + 1$ occurs in a row above i in T. Let

$$(1.54) \qquad \text{maj}(T) = \sum i$$

$$(1.55) \qquad \text{comaj}(T) = \sum |\lambda| - i,$$

where the sums are over the descents of T. Then [**Sta99**, p.363]

$$(1.56) \qquad s_\lambda(1, q, q^2, \ldots,) = \frac{1}{(q)_n} \sum_{T \in SYT(\lambda)} q^{\text{maj}(T)}$$

$$= \frac{1}{(q)_n} \sum_{T \in SYT(\lambda)} q^{\text{comaj}(T)}.$$

The RSK Algorithm

In this subsection we overview some of the basic properties of the famous Robinson-Schensted-Knuth (RSK) algorithm, which gives a bijection between certain two-line arrays of positive integers and pairs of SSYT (P,Q) of the same shape. A more detailed discussion of the RSK algorithm can be found in [**Sta99**].

One of the key components of the algorithm is *row insertion*, which takes as input an arbitrary positive integer k and a SSYT P and outputs a SSYT denoted by $P \leftarrow k$. Row insertion begins by comparing k with the numbers in the bottom row (i.e. row 1) of P. If there are no entries in row 1 which are larger than k, then we form $P \leftarrow k$ by adjoining k to the end of row 1 of P and we are done. Otherwise, we replace (or "bump") the leftmost number in row 1 which is larger than k (say this number is m) by k, and then insert m into row 2 in the same way. That is, if there are no numbers in row 2 larger than m we adjoin m to the end of row 2, otherwise we replace the leftmost number larger than m by m and then insert this number into row 3, etc. It is fairly easy to see that we eventually end up adjoining some number to the end of a row, and so row insertion always terminates, We call this row the terminal row. The output $P \leftarrow k$ is always a SSYT. See Figure 4 for an example.

FIGURE 4. Row insertion of 3 into the tableau on the left results in the 3 bumping the 4 in square $(2,1)$, which bumps the 5 in square $(2,2)$, which is adjoined to the end of row 3 (the terminal row) resulting in the tableau on the right.

Now start with a $2 \times n$ matrix A of positive integers, where the top row is monotone nondecreasing, i.e. $a_{1,i} \leq a_{1,i+1}$. Also assume that below equal entries in the top row, the bottom row is nondecreasing, i.e. that

$$(1.57) \qquad a_{1,i} = a_{1,i+1} \implies a_{2,i} \leq a_{2,i+1}.$$

We call such a matrix an *ordered two-line array*, and let A_1 denote the word $a_{11}a_{12}\cdots a_{1n}$ and A_2 the word $a_{21}a_{22}\cdots a_{2n}$. Given such an A, form a sequence of SSYT $P^{(j)}$ by initializing $P^{(0)} = \emptyset$, then for $1 \leq j \leq n$, let

$$(1.58) \qquad P^{(j)} = P^{(j-1)} \leftarrow a_{2,j}.$$

As we form the $P^{(j)}$, we also form a complementary sequence of SSYT $Q^{(j)}$ by initializing $Q^{(0)} = \emptyset$, and for $1 \leq j \leq n$, we create $Q^{(j)}$ by adjoining $a_{1,j}$ to the end of row r_j of $Q^{(j-1)}$, where r_j is the terminal row of $P^{(j-1)} \leftarrow a_{2,j}$. Finally, we set $P = P^{(n)}$ (the "insertion tableau") and $Q = Q^{(n)}$ (the "recording tableau"). If $A_1 = 112445$ and $A_2 = 331251$ the corresponding P and Q tableau are given in Figure 5.

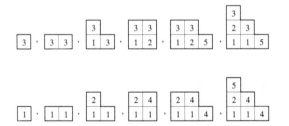

FIGURE 5. On the top, the sequence $P^{(j)}$ and on the bottom, the sequence $Q^{(j)}$, for the ordered two-line array A with $A_1 = 112445$ and $A_2 = 331251$.

THEOREM 1.23. [**Rob38**], [**Sch61**], [**Knu70**] *There is a bijection between ordered two-line arrays A and pairs of SSYT (P, Q) of the same shape. Under this correspondence, the weight of A_1 equals the weight of Q, the weight of A_2 is the weight of P, and $Des(Q) = Des(A_2)$. If A_1 is the identity permutation $12\cdots n$, then we get a bijection between words w and pairs (P_w, Q_w) of tableau of the same shape, with P_w a SSYT of the same weight as w, and Q a SYT. If the top row is the identity permutation and the bottom row is a permutation β, then we get a bijection between permutations β and pairs (P_β, Q_β) of SYT of the same shape.*

REMARK 1.24. It is an interesting fact that for $\beta \in S_n$,

$$P_\beta = Q_{\beta^{-1}}.$$

Plethystic Notation. Many of the theorems later in the course involving symmetric functions will be expressed in plethystic notation. In this subsection we define this and give several examples in order to acclimate the reader.

Let $E(t_1, t_2, t_3 \ldots)$ be a formal series of rational functions in the parameters $t_1, t_2, \ldots$. We define the plethystic substitution of E into p_k, denoted $p_k[E]$, by

(1.59) $$p_k[E] = E(t_1^k, t_2^k, \ldots).$$

Note the square "plethystic" brackets around E - this is to distinguish $p_k[E]$ from the ordinary kth power sum in a set of variables E, which we have already defined as $p_k(E)$. One thing we need to emphasize is that any minus signs occurring in the definition of E are left as is when replacing the t_i by t_i^k.

EXAMPLE 1.25.   (1) $p_k[X] = p_k(X)$. As it is, $p_k[X]$ makes no sense since X indicates a set of variables, not a formal series of rational functions. However, it is traditional to adopt the convention that inside plethystic brackets a capital letter, say Z, stands for $p_1(Z) = z_1 + z_2 + \ldots$.

(2) For z a real parameter,

$$p_k[zX] = z^k p_k[X]$$

(3)

$$p_k[X(1-t)] = \sum_i x_i^k(1 - t^k)$$

(4)

$$p_k[X - Y] = \sum_i x_i^k - y_i^k = p_k[X] - p_k[Y]$$

(5)

$$p_k\left[\frac{X}{1-q}\right] = \sum_i \frac{x_i^k}{1-q^k}$$

(6)

$$p_k\left[\frac{X(1-z)}{1-q}\right] = \sum_i \frac{x_i^k(1-z^k)}{1-q^k}.$$

Note that (4) and (6) are consistent in that

$$p_k\left[\frac{X(1-z)}{1-q}\right] = p_k\left[\frac{X - Xz}{1-q}\right] =$$

$$p_k\left[\frac{X}{1-q} - \frac{Xz}{1-q}\right] = \sum_i \frac{x_i^k}{1-q^k} - \sum_i \frac{z^k x_i^k}{1-q^k}$$

which reduces to $p_k[X]$ when $z = q$, as it should.

Let $Z = (-x_1, -x_2, \ldots)$. Note that $p_k(Z) = \sum_i(-1)^k x_i^k$, which is different from $p_k[-X]$. Thus we need a special notation for the case where we wish to replace variables by their negatives inside plethystic brackets. We use the ϵ symbol to denote this, i.e.

$$(1.60) \qquad p_k[\epsilon X] = \sum_i (-1)^k x_i^k.$$

We now extend the definition of plethystic substitution of E into f for an arbitrary $f \in \Lambda$ by first expressing f as a polynomial in the p_k, say $f = \sum_\lambda c_\lambda p_\lambda$ for constants c_λ, then defining $f[E]$ as

$$(1.61) \qquad f[E] = \sum_\lambda c_\lambda \prod_i p_{\lambda_i}[E].$$

EXAMPLE 1.26. For any $f \in \Lambda$,

$$(1.62) \qquad \omega(f(X)) = f[-\epsilon X].$$

We now derive some very useful identities involving plethysm.

THEOREM 1.27. *The "addition formulas". Let $E = E(t_1, t_2, \ldots)$ and $F = F(w_1, w_2, \ldots)$ be two formal series of rational terms in their indeterminates. Then*

$$(1.63) \qquad e_n[E - F] = \sum_{k=0}^{n} e_k[E] e_{n-k}[-F]$$

$$(1.64) \qquad e_n[E + F] = \sum_{k=0}^{n} e_k[E] e_{n-k}[F]$$

$$(1.65) \qquad h_n[E - F] = \sum_{k=0}^{n} h_k[E] h_{n-k}[-F]$$

$$(1.66) \qquad h_n[E + F] = \sum_{k=0}^{n} h_k[E] h_{n-k}[F].$$

We also have the Cauchy identities

$$(1.67) \qquad e_n[EF] = \sum_{\lambda \in Par(n)} s_\lambda[E] s_{\lambda'}[F]$$

$$(1.68) \qquad h_n[EF] = \sum_{\lambda \in Par(n)} s_\lambda[E] s_\lambda[F].$$

PROOF. By definition of $p_k[E - F]$ and Exercise 1.14,

$$e_n[E - F] = \sum_{\lambda \in \mathrm{Par}(n)} \frac{(-1)^{n - \ell(\lambda)}}{z_\lambda} \prod_m (E(t_1^{\lambda_m}, \ldots) - F(w_1^{\lambda_m}, \ldots))$$

$$= \sum_{\lambda \in \mathrm{Par}(n)} \frac{(-1)^{n - \ell(\lambda)}}{z_\lambda} \prod_i (E(t_1^i, \ldots) - F(w_1^i, \ldots))^{n_i}$$

$$= \sum_{\lambda \in \mathrm{Par}(n)} \frac{(-1)^{n - \ell(\lambda)}}{\prod_i i^{n_i} n_i!} \prod_i \sum_{r_i = 0}^{n_i} \frac{n_i!}{r_i!(n_i - r_i)!} (E(t_1^i, \ldots))^{r_i} (-F(w_1^i, \ldots))^{n_i - r_i}$$

$$= \sum_{\lambda \in \mathrm{Par}(n)} (-1)^{n - \ell(\lambda)}$$

$$\times \prod_i \sum_{r_i = 0}^{n_i} \frac{1}{i^{r_i} r_i! i^{n_i - r_i} (n_i - r_i)!} (E(t_1^i, \ldots))^{r_i} (-F(w_1^i, \ldots))^{n_i - r_i}$$

$$= \sum_{k=0}^{n} \sum_{\substack{\beta \in \mathrm{Par}(k) \\ (n_i(\beta) = r_i)}} \frac{(-1)^{k - \ell(\beta)}}{z_\beta} \prod_m E(t_1^{\beta_m}, \ldots)$$

$$\times \sum_{\substack{\zeta \in \mathrm{Par}(n-k) \\ (n_i(\zeta) = n_i - r_i)}} \frac{(-1)^{n - k - \ell(\zeta)}}{z_\zeta} \prod_m (-F(w_1^{\zeta_m}, \ldots))$$

$$= \sum_{k=0}^{n} e_k[E] e_{n-k}[-F].$$

Minor modifications to the above sequence of steps also proves (1.64), (1.65) and (1.66).

By definition,

$$(1.69) \qquad p_k[EF] = E(t_1^k, \ldots) F(w_1^k, \ldots)$$

so for any $\lambda \in \mathrm{Par}$,

$$(1.70) \qquad p_\lambda[EF] = \prod_i E(t_1^{\lambda_i}, \ldots) F(w_1^{\lambda_i}, \ldots) = p_\lambda[E] p_\lambda[F].$$

Using this both sides of (1.67) and (1.68) can be viewed as polynomial identities in the $p_k[E]$ and $p_k[F]$. Thus they are true if they are true when we replace $p_k[E]$ by $p_k(X)$ and $p_k[F]$ by $p_k(Y)$, in which case they reduce to the well-known non-plethystic Cauchy identities. $\square$

EXAMPLE 1.28. For $\mu \vdash n$, the Hall-Littlewood symmetric function $P_\mu(X; q)$ can be defined as

$$(1.71) \qquad P_\mu(X; q) = \sum_{\lambda \vdash n} \frac{K_{\lambda, \mu}(q)}{\prod_i (q)_{n_i(\lambda)}} s_\lambda[X(1 - q)].$$

(It is well-known that the $s_\lambda[X(1 - q)]$ form a basis for the ring of symmetric functions with coefficients in $\mathbb{Q}(q)$, the family of rational functions in q, and so do the $P_\mu(X; q)$.) Let $\lambda \vdash n$, and z a real parameter. Then $s_\lambda[1 - z] = 0$ if λ is not a "hook" (a hook shape is where $\lambda_2 \leq 1$), in fact

$$(1.72) \qquad s_\lambda[1-z] = \begin{cases} (-z)^r(1-z) & \text{if } \lambda = (n-r, 1^r), \quad 0 \leq r \leq n-1 \\ 0 & \text{else.} \end{cases}$$

Representation Theory

Let G be a finite group. A representation of G is a set of square matrices $\{M(g), g \in G\}$ with the property that

$$(1.73) \qquad M(g)M(h) = M(gh) \quad \forall g, h \in G.$$

On the left-hand-side of (1.73) we are using ordinary matrix multiplication, and on the right-hand-side, to define gh, multiplication in G. The number of rows of a given $M(g)$ is called the dimension of the representation.

An *action* of G on a set S is a map from $G \times S \to S$, denoted by $g(s)$ for $g \in G$, $s \in S$, which satisfies

$$(1.74) \qquad g(h(s)) = (gh)(s) \quad \forall g, h \in G, s \in S.$$

Let V be a finite dimensional $\mathbb{C}$ vector space, with basis $w_1, w_2, \ldots w_n$. Any action of G on V makes V into a $\mathbb{C}G$ module. A module is called *irreducible* if it has no submodules other than $\{0\}$ and itself. Maschke's theorem [**JL01**] says that every V can be expressed as a direct sum of irreducible submodules.

If we form a matrix $M(g)$ whose ith row consists of the coefficients of the w_j when expanding $g(w_i)$ in the w basis, then $\{M(g), g \in G\}$ is a representation of G. In general $\{M(g), g \in G\}$ will depend on the choice of basis, but the trace of the matrices will not. The trace of the matrix $M(g)$ is called the character of the module (under the given action), which we denote char(V). If $V = \bigoplus_{j=1}^d V_j$, where each V_j is irreducible, then the basis can be ordered so that the matrix M will be in block form, where the sizes of the blocks are the dimensions of the V_j. Clearly char$(V) = \sum_{j=1}^d$ char(V_j). It turns out that there are only a certain number of possible functions which occur as characters of irreducible modules, namely one for each conjugacy class of G. These are called the irreducible characters of G.

In the case $G = S_n$, the conjugacy classes are in one-to-one correspondence with partitions $\lambda \in \text{Par}(n)$, and the irreducible characters are denoted $\chi^\lambda(\sigma)$. The dimension of a given V_j with character χ^λ is known to always be f^λ. The value of a given $\chi^\lambda(\sigma)$ depends only on the conjugacy class of σ. For the symmetric group the conjugacy class of an element is determined by rearranging the lengths of the disjoint cycles of σ into nonincreasing order to form a partition called the cycle-type of σ. Thus we can talk about $\chi^\lambda(\beta)$, which means the value of χ^λ at any permutation of cycle type β. For example, $\chi^{(n)}(\beta) = 1$ for all $\beta \vdash n$, so $\chi^{(n)}$ is called the trivial character. Also, $\chi^{1^n}(\beta) = (-1)^{n-\ell(\beta)}$ for all $\beta \vdash n$, so χ^{1^n} is called the sign character, since $(-1)^{n-\ell(\beta)}$ is the sign of any permutation of cycle type β.

One reason Schur functions are important in representation theory is the following.

THEOREM 1.29. *When expanding the s_λ into the p_λ basis, the coefficients are the χ^λ. To be exact*

$$p_\mu = \sum_{\lambda \vdash n} \chi^\lambda(\mu) s_\lambda$$

$$s_\lambda = \sum_{\mu \vdash n} z_\mu^{-1} \chi^\lambda(\mu) p_\mu.$$

Let $\mathbb{C}[X_n] = \mathbb{C}[x_1, \ldots, x_n]$. Given $f(x_1, \ldots, x_n) \in \mathbb{C}[X_n]$ and $\sigma \in S_n$, then

$$(1.75) \qquad\qquad \sigma f = f(x_{\sigma_1}, \ldots, x_{\sigma_n})$$

defines an action of S_n on $\mathbb{C}[X_n]$. We should mention that here we are viewing the permutation σ in a different fashion than when discussing permutation statistics, since here σ is sending the variable in the σ_1st coordinate to the first coordinate, i.e.

$$(1.76) \qquad\qquad \sigma = \begin{pmatrix} \cdots & \sigma_1 & \cdots \\ \cdots & 1 & \cdots \end{pmatrix},$$

which is σ^{-1} in our previous notation. To verify that (1.75) defines an action, note that in this context $\beta(\sigma(f)) = f(x_{\sigma_{\beta_1}}, \ldots)$ while

$$(1.77) \qquad \beta\sigma = \begin{pmatrix} \cdots & \beta_1 & \cdots \\ \cdots & 1 & \cdots \end{pmatrix} \begin{pmatrix} \cdots & \sigma_{\beta_1} & \cdots \\ \cdots & \beta_1 & \cdots \end{pmatrix} = \begin{pmatrix} \cdots & \sigma_{\beta_1} & \cdots \\ \cdots & 1 & \cdots \end{pmatrix}.$$

Let V be a subspace of $\mathbb{C}[X_n]$. Then

$$(1.78) \qquad\qquad V = \sum_{i=0}^{\infty} V^{(i)},$$

where $V^{(i)}$ is the subspace consisting of all elements of V of homogeneous degree i in the x_j. Each $V^{(i)}$ is finite dimensional, and this gives a "grading" of the space V. We define the *Hilbert series* $\mathcal{H}(V; q)$ of V to be the sum

$$(1.79) \qquad\qquad \mathcal{H}(V; q) = \sum_{i=0}^{\infty} q^i \dim(V^{(i)}),$$

where dim indicates the dimension as a $\mathbb{C}$ vector space.

Assume that V is a subspace of $\mathbb{C}[X_n]$ fixed by the S_n action. We define the *Frobenius series* $\mathcal{F}(V; q)$ of V to be the symmetric function

$$(1.80) \qquad\qquad \sum_{i=0}^{\infty} q^i \sum_{\lambda \in \mathrm{Par}(i)} \mathrm{Mult}(\chi^\lambda, V^{(i)}) s_\lambda,$$

where $\mathrm{Mult}(\chi^\lambda, V^{(i)})$ is the multiplicity of the irreducible character χ^λ in the character of $V^{(i)}$ under the action. In other words, if we decompose $V^{(i)}$ into irreducible S_n-submodules, $\mathrm{Mult}(\chi^\lambda, V^{(i)})$ is the number of these submodules whose trace equals χ^λ.

A polynomial in $\mathbb{C}[X_n]$ is *alternating*, or an *alternate*, if

$$(1.81) \qquad\qquad \sigma f = (-1)^{\mathrm{inv}(\sigma)} f \quad \forall \sigma \in S_n.$$

The set of alternates in V forms a subspace called the subspace of alternates, or anti-symmetric elements, denoted V^ϵ. This is also an S_n-submodule of V.

PROPOSITION 1.29.1. *The Hilbert series of V^{ϵ} equals the coefficient of s_{1^n} in the Frobenius series of V, i.e.*

$$(1.82) \qquad \mathcal{H}(V^{\epsilon}; q) = \langle \mathcal{F}(V; q), s_{1^n} \rangle.$$

PROOF. Let B be a basis for $V^{(i)}$ with the property that the matrices $M(\sigma)$ are in block form. Then $b \in B$ is also in $V^{\epsilon(i)}$ if and only if the row of $M(\sigma)$ corresponding to b has entries $(-1)^{\text{inv}(\sigma)}$ on the diagonal and 0's elsewhere, i.e. is a block corresponding to χ^{1^n}. Thus

$$(1.83) \qquad \langle \mathcal{F}(V; q), s_{1^n} \rangle = \sum_{i=0}^{\infty} q^i \dim(V^{\epsilon(i)}) = \mathcal{H}(V^{\epsilon}; q).$$

$\square$

REMARK 1.30. Since the dimension of the representation corresponding to χ^{λ} equals f^{λ}, which by (1.46) equals $< s_{\lambda}, h_{1^n} >$, we have

$$(1.84) \qquad \langle \mathcal{F}(V; q), h_{1^n} \rangle = \mathcal{H}(V; q).$$

EXAMPLE 1.31. Since a basis for $\mathbb{C}[X_n]$ can be obtained by taking all possible monomials in the x_i,

$$(1.85) \qquad \mathcal{H}(\mathbb{C}[X_n]; q) = (1 - q)^{-n}.$$

Taking into account the S_n-action, it is known that

$$(1.86) \qquad \mathcal{F}(\mathbb{C}[X_n]; q) = \sum_{\lambda \in \text{Par}(n)} s_{\lambda} \frac{\sum_{T \in SYT(\lambda)} q^{\text{maj}(T)}}{(q)_n}$$

$$= \sum_{\lambda \in \text{Par}(n)} s_{\lambda} s_{\lambda}(1, q, q^2, \ldots) = \prod_{i,j} \frac{1}{(1 - q^i x_j z)} \Big|_{z^n}.$$

The Ring of Coinvariants. The set of symmetric polynomials in the x_i, denoted $\mathbb{C}[X_n]^{S_n}$, which is generated by $1, e_1, \ldots e_n$, is called the *ring of invariants*. The quotient ring $R_n = \mathbb{C}[x_1, \ldots, x_n] / < e_1, e_2, \ldots, e_n >$, or equivalently $\mathbb{C}[x_1, \ldots, x_n] / < p_1, p_2, \ldots, p_n >$, obtained by moding out by the ideal generated by all symmetric polynomials of positive degree is known as the *ring of coinvariants*. It is known that R_n is finite dimensional as a $\mathbb{C}$-vector space, with $\dim(R_n) = n!$, and more generally that

$$(1.87) \qquad \mathcal{H}(R_n; q) = [n]!.$$

E. Artin [**Art42**] derived a specific basis for R_n, namely

$$(1.88) \qquad \{ \prod_{1 \leq i \leq n} x_i^{\alpha_i}, 0 \leq \alpha_i \leq i - 1 \}.$$

Also,

$$(1.89) \qquad \mathcal{F}(R_n; q) = \sum_{\lambda \in \text{Par}(n)} s_{\lambda} \sum_{T \in SYT(\lambda)} q^{\text{maj}(T)},$$

a result that Stanley [**Sta79**], [**Sta03**] attributes to unpublished work of Lusztig. This shows the Frobenius series of R_n is $(q)_n$ times the Frobenius series of $\mathbb{C}[X_n]$.

Let

$$\Delta = \det \begin{bmatrix} 1 & x_1 & \ldots & x_1^{n-1} \\ 1 & x_2 & \ldots & x_2^{n-1} \\ & & \vdots & \\ 1 & x_n & \ldots & x_n^{n-1} \end{bmatrix} = \prod_{1 \leq i < j \leq n} (x_j - x_i)$$

be the Vandermonde determinant. The *space of harmonics* H_n can be defined as the $\mathbb{C}$ vector space spanned by V_n and its partial derivatives of all orders. Haiman [**Hai94**] provides a detailed proof that H_n is isomorphic to R_n as an S_n module, and notes that an explicit isomorphism α is obtained by letting $\alpha(h), h \in H_n$, be the element of $\mathbb{C}[X_n]$ represented modulo $< e_1, \ldots, e_n >$ by h. Thus $\dim(H_n) = n!$ and moreover the character of H_n under the S_n-action is given by (1.89). He also argues that (1.89) follows immediately from (1.86) and the fact that H_n generates $\mathbb{C}[X_n]$ as a free module over $\mathbb{C}[X_n]^{S_n}$.

CHAPTER 2

Macdonald Polynomials and the Space of Diagonal Harmonics

Kadell and Macdonald's Generalizations of Selberg's Integral

The following result of A. Selberg, known as Selberg's integral [**Sel44**], [**AAR99**] has a number of applications to hypergeometric functions, orthogonal polynomials, and algebraic combinatorics.

THEOREM 2.1. *For n a positive integer and $k, a, b \in \mathbb{C}$ with $\Re a > 0$, $\Re b > 0$, and $\Re k > -min\{1/n, (\Re a)/(n-1), (\Re b)/(n-1)\}$,*

$$(2.1) \qquad \int_{(0,1)^n} | \prod_{1 \le i < j \le n} (x_i - x_j)|^{2k} \prod_{i=1}^{n} x_i^{a-1}(1-x_i)^{b-1} dx_1 \cdots dx_n$$

$$= \prod_{i=1}^{n} \frac{\Gamma(a+(i-1)k)\Gamma(b+(i-1)k)\Gamma(ik+1)}{\Gamma(a+b+(n+i-2)k)\Gamma(k+1)},$$

where $\Gamma(z)$ is the gamma function.

During the 1980's a number of extensions of Selberg's integral were found. Askey obtained a q-analogue of the integral [**Ask80**], while other generalizations involved the insertion of symmetric functions in the x_i into the integrand of (2.1) (see for example [**Sta89**]). One of these extensions, due to Kadell, involved inserting symmetric functions depending on a partition, a set of variables X_n and another parameter. They are now known as Jack symmetric functions since they were first studied by H. Jack [**Jac70**].

For a few partitions of small size Kadell was able to show that a q-analogue of the Jack symmetric functions existed which featured in a q-analogue of his extension of Selberg's integral. He posed the problem of finding q-analogues of these polynomials for all partitions [**Kad88**]. This was solved soon afterwords by Macdonald [**Mac88**], and these q-analogues of Jack symmetric functions are now called Macdonald polynomials, denoted $P_\lambda(X; q, t)$. A brief discussion of their connection to Kadell's work can also be found in [**Mac95**, p.387]. The $P_\lambda(X; q, t)$ are symmetric functions with coefficients in $\mathbb{Q}(q, t)$, the family of rational functions in q and t. If we let $q = t^\alpha$, divide by $(1-t)^{|\lambda|}$ and let $t \to 1^-$ in the P_λ we get the Jack symmetric functions with parameter α. Many other symmetric functions of interest are also limiting or special cases of the $P_\lambda(X; q, t)$, and their introduction was a major breakthrough in algebraic combinatorics and special functions. More recently they have also become important in other areas such as algebraic geometry and commutative algebra.

Here is Macdonald's construction of the $P_\lambda(X; q, t)$. The best reference for their basic properties is [**Mac95**, Chap. 6]. The definition involves the following standard partial order on partitions $\lambda, \mu \in \mathrm{Par}(n)$, called *dominance*.

$$(2.2) \qquad \lambda \geq \mu \iff \sum_{j=1}^{\min(i,\ell(\lambda))} \lambda_j \geq \sum_{j=1}^{\min(i,\ell(\mu))} \mu_j \quad \text{for } i \geq 1.$$

THEOREM 2.2. *Define a q, t extension of the Hall scalar product by*

$$(2.3) \qquad \langle p_\lambda, p_\mu \rangle_{q,t} = \chi(\lambda = \mu) z_\lambda \prod_{i=1}^{\ell(\lambda)} \frac{1 - q^{\lambda_i}}{1 - t^{\lambda_i}}.$$

Then the following conditions uniquely define a family of symmetric functions $\{P_\lambda(X; q, t)\}, \lambda \in Par(n)$ with coefficients in $\mathbb{Q}(q, t)$.

-

$$(2.4) \qquad P_\lambda = \sum_{\mu \leq \lambda} c_{\lambda,\mu} m_\mu,$$

  where $c_{\lambda,\mu} \in \mathbb{Q}(q, t)$ and $c_{\lambda,\lambda} = 1$.

-

$$(2.5) \qquad \langle P_\lambda, P_\mu \rangle_{q,t} = 0 \quad \text{if } \lambda \neq \mu.$$

REMARK 2.3. Since the q, t extension of the Hall scalar product reduces to the ordinary Hall scalar product when $q = t$, it is clear that $P_\lambda(X; q, q) = s_\lambda(X)$.

We can now state Macdonald's q-analogue of Kadell's generalization of Selberg's integral [**Mac95**, Ex. 3, p. 374].

THEOREM 2.4.

$$\frac{1}{n!} \int_{(0,1)^n} P_\lambda(X; q, t) \prod_{1 \leq i < j \leq n} \prod_{r=0}^{k-1} (x_i - q^r x_j)(x_i - q^{-r} x_j)$$

$$\times \prod_{i=1}^{n} x_i^{a-1} (x_i; q)_{b-1} d_q x_1 \cdots d_q x_n$$

$$= q^F \prod_{i=1}^{n} \frac{\Gamma_q(\lambda_i + a + (i-1)k)\Gamma_q(b + (i-1)k)}{\Gamma_q(\lambda_i + a + b + (n+i-2)k)} \prod_{1 \leq i < j \leq n} \frac{\Gamma_q(\lambda_i - \lambda_j + (j-i+1)k)}{\Gamma_q(\lambda_i - \lambda_j + (j-i)k)}$$

where $k \in \mathbb{N}$, $F = kn(\lambda) + kan(n-1)/2 + k^2 n(n-1)(n-2)/3$, $t = q^k$,

$$(2.6) \qquad \Gamma_q(z) = (1-q)^{1-z}(q; q)_\infty/(q^z; q)_\infty$$

is the q-gamma function and

$$(2.7) \qquad \int_0^1 f(x) d_q x = \sum_{i=0}^{\infty} f(q^i)(q^i - q^{i+1})$$

is the q-integral.

EXERCISE 2.5. Show that for functions f which are continuous on $[0, 1]$, the limit as $q \to 1^-$ of the q-integral equals $\int_0^1 f(x) dx$.

Given a cell $x \in \lambda$, let the arm $a = a(x)$, leg $l = l(x)$, coarm $a' = a'(x)$, and coleg $l' = l'(x)$ be the number of cells strictly between x and the border of λ in the E, S, W and N directions, respectively, as in Figure 1. Also, define

$$(2.8) \qquad B_\mu = B_\mu(q,t) = \sum_{x\in\mu}' q^{a'} t^{l'}, \quad \Pi_\mu = \Pi_\mu(q,t) = \prod_{x\in\mu}' (1 - q^{a'} t^{l'}),$$

where a prime symbol $'$ above a product or a sum over cells of a partition μ indicates we ignore the corner $(1,1)$ cell, and $B_\emptyset = 0$, $\Pi_\emptyset = 1$. For example, $B_{(2,2,1)} = 1 + q + t + qt + t^2$ and $\Pi_{(2,2,1)} = (1-q)(1-t)(1-qt)(1-t^2)$. Note that

$$(2.9) \qquad n(\mu) = \sum_{x\in\mu} l' = \sum_{x\in\mu} l.$$

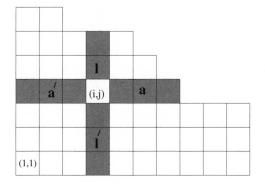

FIGURE 1. The arm a, coarm a', leg l and coleg l' of a cell.

Here are some basic results of Macdonald on the P_λ. Property (2) will be particularly useful to us.

THEOREM 2.6. *let $\lambda, \mu \in Par$.*

(1) *Let z be a real parameter. Then*

$$(2.10) \qquad P_\lambda\left[\frac{1-z}{1-t}; q, t\right] = \prod_{x\in\lambda} \frac{t^{l'} - q^{a'} z}{1 - q^a t^{l+1}}.$$

(2) *(Koornwinder-Macdonald Reciprocity). Assume $n \geq \max(\ell(\lambda), \ell(\mu))$. Then*

$$(2.11) \qquad \frac{\prod_{x\in\mu}(1 - q^a t^{l+1}) P_\lambda\left[\sum_{i=1}^n t^{n-i} q^{\lambda_i}; q, t\right]}{\prod_{x\in\mu}(t^{l'} - q^{a'} t^n)}$$

is symmetric in μ, λ, where as usual we let $\mu_i = 0$ for $i > \ell(\mu)$, $\lambda_i = 0$ for $i > \ell(\lambda)$.

(3) *For any two sets of variables X, Y,*

$$(2.12) \qquad h_n\left[XY\frac{1-t}{1-q}\right] = \sum_{\lambda\vdash n} \frac{\prod_{x\in\lambda}(1 - q^a t^{l+1})}{\prod_{x\in\lambda}(1 - q^{a+1} t^l)} P_\lambda(X; q, t) P_\lambda(Y; q, t)$$

$$(2.13) \qquad e_n[XY] = \sum_{\lambda\vdash n} P_\lambda(X; q, t) P_{\lambda'}(Y; t, q).$$

(4) *Define a q,t version of the involution ω as follows.*

$$(2.14) \qquad \omega_{q,t} p_k[X] = (-1)^{k-1} \frac{1-q^k}{1-t^k} p_k[X].$$

Note this implies

$$(2.15) \qquad \omega_{q,t} p_k \left[X \frac{1-t}{1-q} \right] = (-1)^{k-1} p_k[X].$$

Then

$$(2.16) \qquad \omega_{q,t} P_\lambda[X;q,t] = \prod_{x \in \lambda'} \frac{1-q^{l+1}t^a}{1-q^l t^{a+1}} P_{\lambda'}[X;t,q].$$

EXERCISE 2.7. Show (2.13) can be obtained from (2.12) and (2.16) by applying $\omega_{q,t}$ to the Y set of variables.

The q,t-Kostka Polynomials

Macdonald found that the $P_\lambda(X;q,t)$ have a very mysterious property. Let $J_\mu[X;q,t]$ denote the so-called Macdonald integral form, defined as

$$(2.17) \qquad J_\mu(X;q,t) = \prod_{x \in \mu} (1 - q^a t^{l+1}) P_\mu(X;q,t).$$

Now expand J_μ in terms of the $s_\lambda[X(1-t)]$;

$$(2.18) \qquad J_\mu(X;q,t) = \sum_{\lambda \vdash |\mu|} K_{\lambda,\mu}(q,t) s_\lambda[X(1-t)]$$

for some $K_{\lambda,\mu}(q,t) \in \mathbb{Q}(q,t)$. Macdonald conjectured that $K_{\lambda,\mu}(q,t) \in \mathbb{N}[q,t]$. This became a famous problem in combinatorics known as Macdonald's positivity conjecture.

From Theorem 2.2 and known properties of the Hall-Littlewood polynomials $P_\lambda(X;q)$, it follows that $P_\lambda(X;0,t) = P_\lambda(X;t)$. Also,

$$(2.19) \qquad K_{\lambda,\mu}(0,t) = K_{\lambda,\mu}(t) = \sum_{T \in SSYT(\lambda,\mu)} t^{\text{charge}(T)}.$$

In addition Macdonald was able to show that

$$(2.20) \qquad K_{\lambda,\mu}(1,1) = K_{\lambda,\mu}.$$

Because of (2.19) and (2.20), the $K_{\lambda,\mu}(q,t)$ are known as the q,t-Kostka polynomials.

In addition to Lascoux and Schützenberger's result that the coefficients of $K_{\lambda,\mu}(t)$ can be interpreted combinatorially in terms of the charge statistic, there are a few known ways to prove they are in $\mathbb{N}[t]$ by interpreting them representation theoretically or geometrically [GP92], [Lus81]. This suggests that the nonnegativity of the $K_{\lambda,\mu}(q,t)$ have a similar interpretation. In the next section we describe a conjecture of Garsia and Haiman of just such an interpretation, which was finally proved by Haiman a few years ago, resolving Macdonald's positivity conjecture after more than ten years of intensive research.

Macdonald also posed a refinement of his positivity conjecture which is still open. Due to (2.19) and (2.20), one could hope to find statistics $\mathrm{qstat}(T, \mu)$ and $\mathrm{tstat}(T, \mu)$ given by some combinatorial rule so that

$$(2.21) \qquad K_{\lambda,\mu}(q, t) = \sum_{T \in SYT(\lambda)} q^{\mathrm{qstat}(T,\mu)} t^{\mathrm{tstat}(T,\mu)}.$$

In Garsia and Haiman's work it is more natural to deal with the polynomials

$$(2.22) \qquad \tilde{K}_{\lambda,\mu} = \tilde{K}_{\lambda,\mu}(q, t) = t^{n(\mu)} K_{\lambda,\mu}(q, 1/t),$$

so

$$(2.23) \qquad \tilde{K}_{\lambda,\mu}(0, t) = \sum_{T \in SSYT(\lambda,\mu)} q^{\mathrm{cocharge}(T)}.$$

Macdonald found a statistical description of the $K_{\lambda,\mu}(q, t)$ whenever $\lambda = (n - k, 1^k)$ is a hook shape [**Mac95**, Ex. 2, p. 362], which, stated in terms of the $\tilde{K}_{\lambda,\mu}$, says

$$(2.24) \qquad \tilde{K}_{(n-k,1^k),\mu} = e_k[B_\mu - 1].$$

For example, $\tilde{K}_{(3,1,1),(2,2,1)} = e_2[q + t + qt + t^2] = qt + q^2 t + 2qt^2 + t^3 + qt^3$. He also found statistical descriptions when either q or t is set equal to 1 [**Mac95**, Ex. 7, p. 365]. To describe it, say we are given a statistic $\mathrm{stat}(T)$ on skew SYT, a SYT T with n cells, and a composition $\alpha = (\alpha_1, \ldots, \alpha_k)$ of n into k parts. Define the α-sectionalization of T to be the set of k skew SYT obtained in the following way. The first element of the set is the portion of T containing the numbers 1 through α_1. The second element of the set is the portion of T containing the numbers $\alpha_1 + 1$ through $\alpha_1 + \alpha_2$, but with α_1 subtracted from each of these numbers, so we end up with a skew SYT of size α_2. In general, the ith element of the set, denoted $T^{(i)}$, is the portion of T containing the numbers $\alpha_1 + \ldots + \alpha_{i-1} + 1$ through $\alpha_1 + \ldots + \alpha_i$, but with $\alpha_1 + \ldots + \alpha_{i-1}$ subtracted from each of these numbers. Then we define the α-sectionalization of $\mathrm{stat}(T)$, denoted $\mathrm{stat}(T, \alpha)$, to be the sum

$$(2.25) \qquad \mathrm{stat}(T, \alpha) = \sum_{i=1}^{k} \mathrm{stat}(T^{(i)}).$$

In the above terminology, Macdonald's formula for the $q = 1$ Kostka numbers can be expressed as

$$(2.26) \qquad \tilde{K}_{\lambda,\mu}(1, t) = \sum_{T \in SYT(\lambda)} t^{\mathrm{comaj}(T,\mu')}.$$

For example, given the tableau T in Figure 2 with $\lambda = (4, 3, 2)$ and (coincidentally) μ also $(4, 3, 2)$, then $\mu' = (3, 3, 2, 1)$ and the values of $\mathrm{comaj}(T, \mu')$ on $T^{(1)}, \ldots, T^{(4)}$ are $1, 2, 1, 0$, respectively, so $\mathrm{comaj}(T, \mu') = 4$.

A combinatorial description of the $\tilde{K}_{\lambda,\mu}$ when μ is a hook was found by Stembridge [**Ste94**]. Given a composition α into k parts, let $\mathrm{rev}(\alpha) = (\alpha_k, \alpha_{k-1}, \ldots, \alpha_1)$. Then if $\mu = (n - k, 1^k)$, Stembridge's result can be expressed as

$$(2.27) \qquad \tilde{K}_{\lambda,\mu} = \sum_{T \in SYT(\lambda)} q^{\mathrm{maj}(T,\mu)} t^{\mathrm{comaj}(T,\mathrm{rev}(\mu'))}.$$

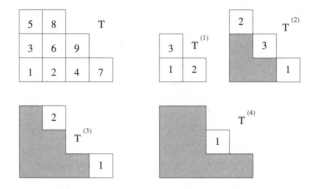

FIGURE 2. The $(3, 3, 2, 1)$-sections of a SYT.

Macdonald obtained the following two symmetry relations.

$$(2.28) \qquad K_{\lambda,\mu}(q,t) = t^{n(\mu)} q^{n(\mu')} K_{\lambda',\mu}(1/q, 1/t)$$

$$(2.29) \qquad K_{\lambda,\mu}(q,t) = K_{\lambda',\mu'}(t,q).$$

Translated into statements about the $\tilde{K}_{\lambda,\mu}$ these become

$$(2.30) \qquad \tilde{K}_{\lambda,\mu}(q,t) = \tilde{K}_{\lambda,\mu'}(t,q)$$

$$(2.31) \qquad \tilde{K}_{\lambda',\mu}(q,t) = t^{n(\mu)} q^{n(\mu')} \tilde{K}_{\lambda,\mu}(1/q, 1/t).$$

Fischel [**Fis95**] first obtained statistics for the case when μ has two columns. Using (2.31) this also implies statistics for the case where μ has two rows. Later Lapointe and Morse [**LM03b**] and Zabrocki [**Zab98**] independently found alternate descriptions of this case, but all of these are rather complicated to state. A simpler description of the two-column case based on the combinatorial formula for Macdonald polynomials is contained in Appendix A.

In 1996 several groups of researchers [**GR98**], [**GT96**], [**KN98**] ,[**Kno97a**], [**Kno97b**], [**LV97**], [**LV98**], [**Sah96**] independently proved that $\tilde{K}_{\lambda,\mu}(q,t)$ is a polynomial with integer coefficients, which itself had been a major unsolved problem since 1988. (The first breakthrough on this problem appears to have been work of Lapointe and Vinet, who in two 1995 CRM preprints [**LV95a**],[**LV95b**] proved the corresponding integrality result for Jack polynomials. This seemed to have the effect of breaking the ice, since it was shortly after this that the proofs of Macdonald integrality were announced. As in the work of Kadell on Selberg's integral, this gives another example of how results in Macdonald theory are often preceeded by results on Jack polynomials.) We should mention that the Macdonald polynomialty result is immediately implied by the combinatorial formula in Appendix A. The paper by Garsia and Remmel also contains a recursive formula for the $\tilde{K}_{\lambda,\mu}(q,t)$ when λ is an augmented hook, i.e. a hook plus the square $(2,2)$. Their formula immediately implies nonnegativity and by iteration could be used to obtain various combinatorial descriptions for this case. In the late 1990's Tesler announced that using plethystic methods he could prove nonnegativity of the case where λ is a "doubly augmented hook", which is an augmented hook plus either the cell $(2,3)$ or $(3,2)$ [**Tes99**].

The Garsia-Haiman Modules and the $n!$-Conjecture

Given any subspace $W \subseteq \mathbb{C}[X_n, Y_n]$, we define the bigraded Hilbert series of W as

$$(2.32) \qquad \mathcal{H}(W; q, t) = \sum_{i,j \geq 0} t^i q^j \dim(W^{(i,j)}),$$

where the subspaces $W^{(i,j)}$ consist of those elements of W of bi-homogeneous degree i in the x variables and j in the y variables, so $W = \bigoplus_{i,j \geq 0} W^{(i,j)}$. Also define the *diagonal action* of S_n on W by

$$(2.33) \qquad \sigma f = f(x_{\sigma_1}, \ldots x_{\sigma_n}, y_{\sigma_1}, \ldots, y_{\sigma_n}) \quad \sigma \in S_n, \, f \in W.$$

Clearly the diagonal action fixes the subspaces $W^{(i,j)}$, so we can define the bigraded Frobenius series of W as

$$(2.34) \qquad \mathcal{F}(W; q, t) = \sum_{i,j \geq 0} t^i q^j \sum_{\lambda \vdash n} s_\lambda \mathrm{Mult}(\chi^\lambda, W^{(i,j)}).$$

Similarly, let W^ϵ be the subspace of alternating elements in W, and

$$(2.35) \qquad \mathcal{H}(W^\epsilon; q, t) = \sum_{i,j \geq 0} t^i q^j \dim(W^{\epsilon(i,j)}).$$

As in the case of subspaces of $\mathbb{C}[X_n]$,

$$(2.36) \qquad \mathcal{H}(W^\epsilon; q, t) = \langle \mathcal{F}(W^\epsilon; q, t), s_{1^n} \rangle .$$

For $\mu \in \mathrm{Par}(n)$, let $(r_1, c_1), \ldots, (r_n, c_n)$ be the $(a'-1, l'-1) = (column, row)$ coordinates of the cells of μ, taken in some arbitrary order, and let

$$(2.37) \qquad \Delta_\mu(X_n, Y_n) = \left| x_i^{r_j - 1} y_i^{c_j - 1} \right|_{i,j=1,n}.$$

For example,

$$(2.38) \qquad \Delta_{(2,2,1)}(X_5, Y_5) = \begin{vmatrix} 1 & y_1 & x_1 & x_1 y_1 & x_1^2 \\ 1 & y_2 & x_2 & x_2 y_2 & x_2^2 \\ 1 & y_3 & x_3 & x_3 y_3 & x_3^2 \\ 1 & y_4 & x_4 & x_4 y_4 & x_4^2 \\ 1 & y_5 & x_5 & x_5 y_5 & x_5^2 \end{vmatrix}.$$

Note that, up to sign, $\Delta_{1^n}(X_n, 0) = \Delta(X_n)$, the Vandermonde determinant.

For $\mu \vdash n$, let $V(\mu)$ denote the linear span of $\Delta_\mu(X_n, Y_n)$ and its partial derivatives of all orders. Note that, although the sign of Δ_μ may depend on the arbitrary ordering of the cells of μ we started with, $V(\mu)$ is independent of this ordering. Garsia and Haiman conjectured [**GH93**] the following result, which was proved by Haiman in 2001 [**Hai01**].

THEOREM 2.8. *For all* $\mu \vdash n$,

$$(2.39) \qquad \mathcal{F}(V(\mu); q, t) = \tilde{H}_\mu,$$

where $\tilde{H}_\mu = \tilde{H}_\mu[X; q, t]$ *is the "modified Macdonald polynomial" defined as*

$$(2.40) \qquad \tilde{H}_\mu = \sum_{\lambda \vdash n} \tilde{K}_{\lambda,\mu}(q, t) s_\lambda.$$

Note that Theorem 2.8 implies $\tilde{K}_{\lambda,\mu}(q, t) \in \mathbb{N}[q, t]$.

COROLLARY 2.8.1. *For all* $\mu \vdash n$, $\dim V(\mu) = n!$.

PROOF. We obtain $\dim(V(\mu))$ from the Frobenius series by replacing each s_λ by f^λ, then letting $q = t = 1$. Thus, using (2.20) we get

$$(2.41) \qquad \dim(V(\mu)) = \sum_{\lambda \vdash n} f^\lambda f^\lambda,$$

which is well known to equal $n!$. $\square$

Theorem 2.8 and Corollary 2.8.1 were together known as the "$n!$ conjecture". Although Corollary 2.8.1 appears to be much weaker, Haiman proved [**Hai99**] in the late 1990's that Corollary 2.8.1 implies Theorem 2.8.

It is clear from the definition of $V(\mu)$ that

$$(2.42) \qquad \mathcal{F}(V(\mu); q, t) = \mathcal{F}(V(\mu'); t, q).$$

Theorem 2.8 thus gives a geometric interpretation to (2.30).

EXAMPLE 2.9. We have

$$
\begin{aligned}
\tilde{H}_{1^n}[X; q, t] &= \mathcal{F}(V(1^n); q, t) \\
&= \mathcal{F}(H_n; t) \\
&= \sum_{\lambda \vdash n} (t; t)_n s_\lambda(1, t, t^2, \ldots,) s_\lambda \\
&= \sum_{\lambda \vdash n} s_\lambda \sum_{T \in SYT(\lambda)} t^{\mathrm{maj}(T)} \\
&= (t; t)_n h_n \left[\frac{X}{1-t} \right],
\end{aligned}
$$

from (1.89), (1.68) and (1.56). By (2.42) we also have

$$(2.43) \qquad \tilde{H}_n = (q)_n h_n \left[\frac{X}{1-q} \right].$$

EXERCISE 2.10. Show (2.24) is equivalent to

$$(2.44) \qquad \left\langle \tilde{H}_\mu, e_{n-d} h_d \right\rangle = e_{n-d}[B_\mu], \qquad \mu \vdash n.$$

Use this to conclude there is one occurrence of the trivial representation in $V(\mu)$, corresponding to $V(\mu)^{(0,0)}$, and one occurrence of the sign representation, corresponding to $V(\mu)^{(n(\mu), n(\mu'))} = \Delta_\mu$.

Before Haiman proved the general case using algebraic geometry, Garsia and Haiman proved the special case of the $n!$ conjecture when μ is a hook by combinatorial methods [**GH96b**]. The case where μ is an augmented hook was proved by Reiner [**Rei96**].

From (2.17) we see that

$$
\begin{aligned}
(2.45) \qquad \tilde{H}_\mu[X; q, t] &= t^{n(\mu)} J_\mu \left[\frac{X}{1 - 1/t}; q, 1/t \right] \\
&= t^{n(\mu)} P_\mu \left[\frac{X}{1 - 1/t}; q, 1/t \right] \prod_{x \in \mu} (1 - q^a t^{l+1}).
\end{aligned}
$$

Macdonald derived formulas for the coefficients in the expansion of $e_k P_\mu(X; q, t)$ and also $h_k \left[X \frac{(1-t)}{1-q} \right] P_\mu(X; q, t)$ in terms of the $P_\lambda(X; q, t)$. These expansions reduce to the classical Pieri formulas for Schur functions discussed in Chapter 1 in the case $t = q$. When expressed in terms of the J_μ, the h_k Pieri rule becomes [**Mac95**, Ex. 4, p.363]

(2.46)

$$h_k \left[X \frac{(1-t)}{1-q} \right] J_\mu = \sum_{\substack{\lambda \in \mathrm{Par} \\ \lambda/\mu \text{ is a horizontal } k\text{-strip}}} \frac{\prod_{x \in \mu} (1 - q^{a_\mu + \chi(x \in B)} t^{l_\mu + \chi(x \notin B)})}{\prod_{x \in \lambda} (1 - q^{a_\lambda + \chi(x \in B)} t^{l_\lambda + \chi(x \notin B)})} J_\lambda,$$

where B is the set of columns which contain a cell of λ/μ, a_μ, l_μ are the values of a, l when the cell is viewed as part of μ, and a_λ, l_λ are the values of a, l when the cell is viewed as part of λ.

EXERCISE 2.11. Show that (2.46), when expressed in terms of the $\tilde{H}_\mu$, becomes

$$(2.47) \quad h_k \left[\frac{X}{1-q} \right] \tilde{H}_\mu = \sum_{\substack{\lambda \in \mathrm{Par} \\ \lambda/\mu \text{ is a horizontal } k\text{-strip}}} \frac{\prod_{x \in \mu} (t^{l_\mu + \chi(x \notin B)} - q^{a_\mu + \chi(x \in B)})}{\prod_{x \in \lambda} (t^{l_\lambda + \chi(x \notin B)} - q^{a_\lambda + \chi(x \in B)})} \tilde{H}_\lambda.$$

The Space of Diagonal Harmonics

Let $p_{h,k}[X_n, Y_n] = \sum_{i=1}^{n} x_i^h y_i^k$, $h, k \in \mathbb{N}$ denote the "polarized power sum". It is known that the set $\{p_{h,k}[X_n, Y_n], h + k \geq 0\}$ generate $\mathbb{C}[X_n, Y_n]^{S_n}$, the ring of invariants under the diagonal action. Thus the natural analog of the quotient ring R_n of coinvariants is the quotient ring DR_n of diagonal coinvariants defined by

$$(2.48) \qquad DR_n = \mathbb{C}[X_n, Y_n] / \left\langle \sum_{i=1}^{n} x_i^h y_i^k, \forall h + k > 0 \right\rangle.$$

By analogy we also define the space of diagonal harmonics DH_n by

$$(2.49) \qquad DH_n = \left\{ f \in \mathbb{C}[X_n, Y_n] : \sum_{i=1}^{n} \frac{\partial^h}{\partial x_i^h} \frac{\partial^k}{\partial y_i^k} f = 0, \forall h + k > 0. \right\}.$$

Many of the properties of H_n and R_n carry over to two sets of variables. For example DH_n is a finite dimensional vector space which is isomorphic to DR_n. The dimension of these spaces turns out to be $(n+1)^{n-1}$, a result which was first conjectured by Haiman [**Hai94**] and proved by him in 2001 [**Hai02**]. His proof uses many of the techniques and results from his proof of the $n!$ conjecture. See [**Sta03**] for a nice expository account of the $n!$ conjecture and the $(n+1)^{n-1}$ theorem.

EXERCISE 2.12.   (1) Show DH_n is closed under partial differentiation.
  (2) For $\mu \vdash n$, prove that $\Delta_\mu \in DH_n$. Conclude that $V(\mu)$ is an S_n-submodule of DH_n.

EXAMPLE 2.13. An explicit basis for DH_2 is given by $\{1, x_2 - x_1, y_2 - y_1\}$. The elements $x_2 - x_1$ and $y_2 - y_1$ form a basis for DH_2^ϵ. Thus

$$(2.50) \qquad \mathcal{F}(DH_2; q, t) = s_2 + (q + t)s_{1^2}.$$

The number $(n+1)^{n-1}$ is known to count some interesting combinatorial structures. For example, it counts the number of rooted, labelled trees on $n+1$ vertices with root node labelled 0. It also counts the number of parking functions on n cars, which can be identified with placements of the numbers 1 through n immediately to the right of N steps in some Dyck path $\pi \in L_{n,n}^+$, with strict decrease down

columns. In Chapter 5 we discuss a conjecture, which is still open, of Haglund and
Loehr which gives a combinatorial description for $\mathcal{H}(DH_n; q, t)$ in terms of statistics
on parking functions [**HL05**].

We let $M = (1 - q)(1 - t)$ and for $\mu \in \mathrm{Par}$,

$$(2.51) \qquad T_\mu = t^{n(\mu)} q^{n(\mu')}, \quad w_\mu = \prod_{x \in \mu} (q^a - t^{l+1})(t^l - q^{a+1}).$$

Haiman derives the $(n+1)^{n-1}$ result as a corollary of the following formula for the
Frobenius series of DH_n.

THEOREM 2.14. *(Haiman, [**Hai02**]).*

$$(2.52) \qquad \mathcal{F}(DH_n; q, t) = \sum_{\mu \vdash n} \frac{T_\mu M \tilde{H}_\mu \Pi_\mu B_\mu}{w_\mu}.$$

Theorem 2.14 was conjectured by Garsia and Haiman in [**GH96a**]. The conjecture was inspired in part by suggestions of C. Procesi.

From (2.44) and the fact that $T_\mu = e_n[B_\mu]$, we have $< \tilde{H}_\mu, s_{1^n} > = T_\mu$. Thus if
we define

$$(2.53) \qquad C_n(q, t) = \sum_{\mu \vdash n} \frac{T_\mu^2 M \Pi_\mu B_\mu}{w_\mu},$$

then by (2.52),

$$(2.54) \qquad \begin{aligned} C_n(q, t) &= \langle \mathcal{F}(DH_n; q, t), s_{1^n} \rangle \\ &= \mathcal{H}(DH_n^\epsilon; q, t). \end{aligned}$$

For instance, from Example 2.13, we have $C_2(q, t) = q + t$. $C_n(q, t)$ is referred to as
the q, t-Catalan sequence, since Garsia and Haiman proved that $C_n(1, 1)$ reduces
to $C_n = \frac{1}{n+1} \binom{2n}{n}$, the nth Catalan number. The C_n have quite a history and arise
very frequently in combinatorics and elsewhere. See [**Sta99**, Ex. 6.19], [**Sta07**] for
over 160 different objects counted by the Catalan numbers.

Haglund introduced a conjectured combinatorial description for $C_n(q, t)$ in
terms of statistics on Dyck paths $\pi \in L_{n,n}^+$ [**Hag03**], which we describe in Chapter
3. Garsia and Haglund then proved this conjecture using plethystic identities for
Macdonald polynomials [**GH01**], [**GH02**]. Later Egge, Haglund, Killpatrick and
Kremer introduced a conjectured combinatorial description for the coefficient of a
hook shape in $\mathcal{F}(DH_n; q, t)$ in terms of statistics on "Schröder" paths, which are
lattice paths consisting of North, East and diagonal D $(1, 1)$ steps [**EHKK03**].
The resulting "(q, t)-Schröder polynomial" is the subject of Chapter 4. In Chapter
7 we discuss Haglund's proof of this conjecture [**Hag03**], which is an extension of
Garsia and Haglund's proof of the combinatorial formula for $C_n(q, t)$.

OPEN PROBLEM 2.15. *Find a combinatorial description of the polynomials*
$< \mathcal{F}(DH_n; q, t), s_\lambda >$ *for general* λ.

We refer to Problem 2.15 as the "fundamental problem", since it, together with
Macdonald's positivity conjecture, has motivated most of the research described in
this book. From the preceding paragraph we note that the fundamental problem
has been solved for hook shapes. We remark that the only known way at present
to prove this result is to use both Theorem 2.14 and plethystic symmetric function

identities. One could eventually hope to find an explicit basis for the irreducible S_n-submodules of DH_n which would provide an immediate solution to the fundamental problem by simply reading off the degrees of appropriately selected basis elements.

In Chapter 6 we discuss a recent conjecture of Haglund, Haiman, Loehr, Remmel and Ulyanov [**HHL$^+$05c**], which gives a conjectured formula for the coefficients in the expansion of the Frobenius series of DH_n into monomial symmetric functions. We will refer to this as the "shuffle conjecture", since it can be described in terms of combinatorial statistics on certain permutations associated to parking functions which are shuffles of blocks of increasing and decreasing sequences. This conjecture includes the conjecture for the Hilbert series of DH_n due to Haglund and Loehr (discussed in Chapter 5) and the formula for hook shapes in terms of Schröder paths as special cases. It doesn't quite imply nonnegativity of the coefficients of the Schur functions, since the expansion of m_λ into Schur functions has some negative coefficients, but it does represent substantial progress on Problem 2.15. In Chapter 6 we discuss several propositions which give supporting evidence for the conjecture.

The Nabla Operator

We begin this section with a slight generalization of the Koornwinder-Macdonald reciprocity formula, in a form which occurs in [**GHT99**].

THEOREM 2.16. *Let* $\mu, \lambda \in Par$, $z \in \mathbb{R}$. *Then*

$$(2.55) \qquad \frac{\tilde{H}_\mu[1 + z(MB_\lambda - 1); q, t]}{\prod_{x \in \mu}(1 - zq^{a'}t^{l'})} = \frac{\tilde{H}_\lambda[1 + z(MB_\mu - 1); q, t]}{\prod_{x \in \lambda}(1 - zq^{a'}t^{l'})}.$$

PROOF. The left-hand-side of (2.55), expressed in terms of the P_λ using (2.45) becomes

$$(2.56) \qquad t^{n(\mu)} \frac{\prod_{x \in \mu}(1 - q^a/t^{l+1})}{\prod_{x \in \mu}(1 - zq^{a'}t^{l'})} P_\mu \left[\frac{1 - z + zMB_\lambda}{1 - 1/t}; q, 1/t \right].$$

Replacing t by $1/t$ in (2.56) we get

$$(2.57) \qquad \frac{\prod_{x \in \mu}(1 - q^a t^{l+1})}{\prod_{x \in \mu}(t^{l'} - zq^{a'})} P_\mu \left[\frac{1 - z}{1 - t} - t^{-1}z(1 - q)B_\lambda(q, 1/t); q, t \right].$$

Now assume $z = t^n$, where $n \geq \max(\ell(\mu), \ell(\lambda))$. Since

$$(2.58) \qquad B_\lambda(q, t) = \sum_{i \geq 1} t^{i-1} \frac{1 - q^{\lambda_i}}{1 - q},$$

eq. (2.57) equals

$$(2.59) \qquad \frac{\prod_{x \in \mu}(1 - q^a t^{l+1})}{\prod_{x \in \mu}(t^{l'} - t^n q^{a'})} P_\mu \left[1 + t + \ldots + t^{n-1} - (\sum_{i=1}^{n} t^{n-i}(1 - q^{\lambda_i})); q, t \right]$$

$$(2.60) \qquad = \frac{\prod_{x \in \mu}(1 - q^a t^{l+1})}{\prod_{x \in \mu}(t^{l'} - t^n q^{a'})} P_\mu \left[\sum_{i=1}^{n} t^{n-i} q^{\lambda_i}; q, t \right]$$

which is symmetric in μ, λ by (2.11). Eq. (2.55) thus follows for $z = t^n$. By cross multiplying, we can rewrite (2.55) as a statement saying two polynomials in z are equal, and two polynomials which agree on infinitely many values $z = t^n$ must be identically equal. $\qquad \square$

REMARK 2.17. If $|\mu|, |\lambda| > 0$, then we can cancel the factor of $1 - z$ in the denominators on both sides of 2.55 and then set $z = 1$ to obtain

$$(2.61) \qquad \frac{\tilde{H}_\mu[MB_\lambda; q, t]}{\Pi_\mu} = \frac{\tilde{H}_\lambda[MB_\mu; q, t]}{\Pi_\lambda}.$$

Another useful special case of 2.55 is $\lambda = \emptyset$, which gives

$$(2.62) \qquad \tilde{H}_\mu[1 - z; q, t] = \prod_{x \in \mu}(1 - zq^{a'}t^{l'}).$$

Let ∇ be the linear operator on symmetric functions which satisfies

$$(2.63) \qquad \nabla \tilde{H}_\mu = T_\mu \tilde{H}_\mu.$$

It turns out that many of the results in Macdonald polynomials and diagonal harmonics can be elegantly expressed in terms of ∇. Some of the basic properties of ∇ were first worked out by F. Bergeron [**Ber96**], and more advanced applications followed in a series of papers by Bergeron, Garsia, Haiman and Tesler [**BGHT99**], [**BG99**], [**GHT99**].

PROPOSITION 2.17.1.

$$(2.64) \qquad \nabla e_n = \sum_{\mu \vdash n} \frac{T_\mu M \tilde{H}_\mu \Pi_\mu B_\mu}{w_\mu}, \qquad n > 0.$$

Hence Theorem 2.14 is equivalent to

$$(2.65) \qquad \mathcal{F}(DH_n; q, t) = \nabla e_n.$$

PROOF. Replacing X by $X/(1 - t)$, Y by $Y/(1 - t)$, and then letting $t = 1/t$ in (2.12) gives

$$(2.66)$$
$$h_n\left[\frac{XY(1 - 1/t)}{(1 - 1/t)(1 - 1/t)(1 - q)}\right]$$
$$= \sum_{\lambda \vdash n} \frac{\prod_{x \in \mu}(1 - q^a t^{-(l+1)})}{\prod_{x \in \mu}(1 - q^{a+1}t^{-l})} P_\lambda(X/(1 - 1/t); q, 1/t) P_\lambda(Y/(1 - 1/t); q, 1/t).$$

Using (2.45) this can be expressed as

$$(2.67) \qquad h_n\left[-t\frac{XY}{M}\right] = \sum_{\mu \vdash n} \frac{\tilde{H}_\mu[X; q, t]\tilde{H}_\mu[Y; q, t] \prod_{x \in \mu}(1 - q^a t^{-(l+1)})}{t^{2n(\mu)} \prod_{x \in \mu}(1 - q^a t^{-(l+1)})^2 \prod_{x \in \mu}(1 - q^{a+1}t^{-l})}.$$

Since $h_n[-X] = (-1)^n e_n[X]$, this is equivalent to

$$(2.68) \qquad e_n\left[\frac{XY}{M}\right] = \sum_{\mu \vdash n} \frac{\tilde{H}_\mu[X; q, t]\tilde{H}_\mu[Y; q, t]}{\prod_{x \in \mu}(t^l - q^{a+1})(t^{l+1} - q^a)(-1)^n}$$

$$(2.69) \qquad = \sum_{\mu \vdash n} \frac{\tilde{H}_\mu[X; q, t]\tilde{H}_\mu[Y; q, t]}{w_\mu}.$$

Now letting $\lambda = 1$ in (2.55), canceling the common factor of $1 - z$ in the denominator of each side, then letting $z = 1$ we get

$$(2.70) \qquad \tilde{H}_\mu[M; q, t] = \prod_{x \in \mu}(1 - q^{a'}t^{l'})\tilde{H}_{(1)}[MB_\mu; q, t].$$

Since $\tilde{H}_{(1)}[X; q, t] = X$, this implies

(2.71) $\tilde{H}_\mu[M; q, t] = \Pi_\mu M B_\mu.$

Letting $Y = M$ in (2.69) and using (2.71) we finally arrive at

(2.72) $(-1)^n e_n = \sum_{\mu \vdash n} \dfrac{M \tilde{H}_\mu \Pi_\mu B_\mu}{w_\mu (-1)^n}.$

Taking ∇ of both sides completes the proof.   □

There is more general form of DH_n studied by Haiman, which depends on a positive integer m, which we denote $DH_n^{(m)}$. For $m = 1$ it reduces to DH_n, and Haiman has proved that the Frobenius series of these spaces can be expressed as $\nabla^m e_n$ [**Hai02**]. As a corollary he proves an earlier conjecture, that the dimension of the subspace of alternants, $DH_n^{\epsilon(m)}$, equals $|L_{nm,n}^+|$, the number of lattice paths in a $nm \times n$ rectangle not going below the line $y = x/m$, which can be viewed as a parameter m Catalan number C_n^m. Most of the lattice-path combinatorics in the $m = 1$ case follows through for general m in an elegant fashion. In particular there are conjectured combinatorial statistics due to Haiman and Loehr [**Hai00a**] [**Loe03**] for $\mathcal{H}(DH_n^{\epsilon(m)})$. This conjecture is still open for all $m > 1$. There is also a version of the shuffle conjecture for these spaces, which we describe at the end of Chapter 6.

EXERCISE 2.18. In this exercise we prove the following alternate version of Koornwinder-Macdonald reciprocity, which is an unpublished result of the author.

(2.73)
$$\dfrac{\tilde{H}_\mu[1 - z - (1 - 1/q)(1 - 1/t)B_\lambda(1/q, 1/t); q, t]}{\prod_{x \in \mu}(1 - zq^{a'}t^{l'})}$$
$$= \dfrac{\tilde{H}_\lambda[1 - z - (1 - 1/q)(1 - 1/t)B_\mu(1/q, 1/t); q, t]}{\prod_{x \in \lambda}(1 - zq^{a'}t^{l'})}.$$

(1) Show that the left-hand-side of 2.73, after expressing it in terms of the the P_λ and replacing q by $1/q$ equals

(2.74) $t^{n(\mu)} \displaystyle\prod_{s \in \mu} \dfrac{1 - 1/q^a t^{l+1}}{1 - zt^{l'}/q^{a'}} P_\mu \left[\dfrac{1 - z}{1 - 1/t} - (1 - q)B_\lambda(q, 1/t); 1/q, 1/t \right].$

(2) Use the symmetry relation $P_\mu(X; 1/q, 1/t) = P_\mu(X; q, t)$ [**Mac95**, p. 324] to show the $z = 1/t^N$ case of (2.74) is equivalent to the $z = t^N$ case of (2.11), where $N \in \mathbb{N}, N \geq \ell(\lambda), \ell(\mu)$. Conclude that (2.73) is true for infinitely many integer values of z. Show that (2.73) can be viewed as a polynomial identity in z, and is hence true for all $z \in \mathbb{C}$.

CHAPTER 3

The q,t-Catalan Numbers

The Bounce Statistic

In this section we give a combinatorial formula for $C_n(q,t)$, the q,t-Catalan number. Our formula involves a new statistic on Dyck paths we call *bounce*.

DEFINITION 3.1. Given $\pi \in L_{n,n}^+$, define the *bounce path* of π to be the path described by the following algorithm.
Start at $(0,0)$ and travel North along π until you encounter the beginning of an E step. Then turn East and travel straight until you hit the diagonal $y = x$. Then turn North and travel straight until you again encounter the beginning of an E step of π, then turn East and travel to the diagonal, etc. Continue in this way until you arrive at (n,n).

We can think of our bounce path as describing the trail of a billiard ball shot North from $(0,0)$, which "bounces" right whenever it encounters a horizontal step and "bounces" up when it encounters the line $y = x$. The bouncing ball will strike the diagonal at places $(0,0), (j_1, j_1), (j_2, j_2), \ldots, (j_{b-1}, j_{b-1}), (j_b, j_b) = (n,n)$. We define the bounce statistic bounce(π) to be the sum

$$(3.1) \qquad \text{bounce}(\pi) = \sum_{i=1}^{b-1} n - j_i,$$

and we call b the number of bounces with j_1 the length of the first bounce, $j_2 - j_1$ the length of the second bounce, etc. The lattice points where the bouncing billiard ball switches from traveling North to East are called the *peaks* of π. The first peak is the peak with smallest y coordinate, the second peak the one with next smallest y coordinate, etc. For example, for the path π in Figure 1, there are 5 bounces of lengths $3, 2, 2, 3, 1$ and bounce$(\pi) = 19$. The first two peaks have coordinates $(0,3)$ and $(3,5)$.

Let

$$(3.2) \qquad F_n(q,t) = \sum_{\pi \in L_{n,n}^+} q^{\text{area}(\pi)} t^{\text{bounce}(\pi)}.$$

THEOREM 3.2.

$$(3.3) \qquad C_n(q,t) = F_n(q,t).$$

Combining Theorems 3.2 and 2.14 we have the following.

COROLLARY 3.2.1.

$$(3.4) \qquad \mathcal{H}(DH_n^\epsilon; q, t) = \sum_{\pi \in L_{n,n}^+} q^{area(\pi)} t^{bounce(\pi)}.$$

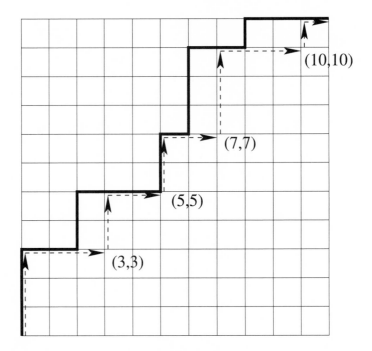

FIGURE 1. The bounce path (dotted line) of a Dyck path (solid line). The bounce statistic equals $11-3+11-5+11-7+11-10 = 8+6+4+1 = 19$.

Theorem 3.2 was first conjectured by Haglund in 2000 [**Hag03**] after a prolonged study of tables of $C_n(q,t)$. It was then proved by Garsia and Haglund [**GH01**], [**GH02**]. At the present time there is no known way of proving Corollary 3.2.1 without using both Theorems 3.2 and 2.14.

The proof of Theorem 3.2 is based on a recursive structure underlying $F_n(q,t)$.

DEFINITION 3.3. Let $L_{n,n}^{+}(k)$ denote the set of all $\pi \in L_{n,n}^{+}$ which begin with exactly k N steps followed by an E step. By convention $L_{0,0}^{+}(k)$ consists of the empty path if $k = 0$ and is empty otherwise. Set

$$(3.5) \qquad F_{n,k}(q,t) = \sum_{\pi \in L_{n,n}^{+}(k)} q^{\text{area}(\pi)} t^{\text{bounce}(\pi)}, \quad F_{n,0} = \chi(n=0).$$

The following recurrence was first obtained in [**Hag03**].

THEOREM 3.4. *For* $1 \le k \le n$,

$$(3.6) \qquad F_{n,k}(q,t) = \sum_{r=0}^{n-k} \begin{bmatrix} r+k-1 \\ r \end{bmatrix} t^{n-k} q^{\binom{k}{2}} F_{n-k,r}(q,t).$$

PROOF. Given $\beta \in L_{n,n}^{+}(k)$, with first bounce k and second bounce say r, then β must pass through the lattice points with coordinates $(1,k)$ and $(k,k+r)$ (the two large dots in Figure 2). Decompose β into two parts, the first part being the portion of β starting at $(0,0)$ and ending at $(k,k+r)$, and the second the portion starting at $(k,k+r)$ and ending at (n,n). If we adjoin a sequence of r N steps to the beginning of the second part, we obtain a path β' in $L^+_{n-k,n-k}(r)$. It is easy to check that $\text{bounce}(\beta) = \text{bounce}(\beta') + n - k$. It remains to relate $\text{area}(\beta)$ and $\text{area}(\beta')$.

Clearly the area inside the triangle below the first bounce step is $\binom{k}{2}$. If we fix β', and let β vary over all paths in $L^+_{n,n}(k)$ which travel through $(1,k)$ and $(k,k+r)$, then the sum of $q^{\text{area}(\beta)}$ will equal

$$(3.7) \qquad q^{\text{area}(\beta')}q^{\binom{k}{2}}\begin{bmatrix} k+r-1 \\ r \end{bmatrix}$$

by (1.6). Thus

$$(3.8) \qquad F_{n,k}(q,t) = \sum_{r=0}^{n-k} \sum_{\beta' \in L^+_{n-k,n-k}(r)} q^{\text{area}(\beta')}t^{\text{bounce}(\beta')}t^{n-k}q^{\binom{k}{2}}\begin{bmatrix} k+r-1 \\ r \end{bmatrix}$$

$$(3.9) \qquad = \sum_{r=0}^{n-k}\begin{bmatrix} r+k-1 \\ r \end{bmatrix}t^{n-k}q^{\binom{k}{2}}F_{n-k,r}(q,t).$$

$\square$

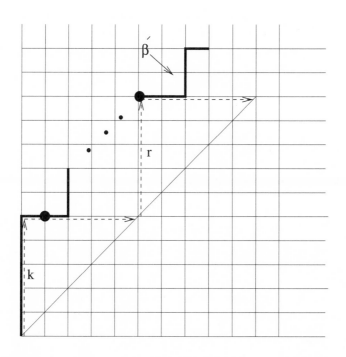

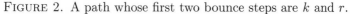

FIGURE 2. A path whose first two bounce steps are k and r.

COROLLARY 3.4.1.
$$(3.10)$$

$$F_n(q,t) = \sum_{b=1}^{n} \sum_{\substack{\alpha_1+\alpha_2+\ldots+\alpha_b=n \\ \alpha_i>0}} t^{\alpha_2+2\alpha_3+\ldots+(b-1)\alpha_b}q^{\sum_{i=1}^{b}\binom{\alpha_i}{2}}\prod_{i=1}^{b-1}\begin{bmatrix} \alpha_i+\alpha_{i+1}-1 \\ \alpha_{i+1} \end{bmatrix},$$

where the inner sum is over all compositions α of n into b positive integers.

PROOF. This follows by iterating the recurrence in Theorem 3.4. The inner term in the sum over b equals the sum of $q^{\text{area}} t^{\text{bounce}}$ over all paths π whose bounce path has b steps of lengths $\alpha_1, \ldots, \alpha_b$. For example, for such a π, the contribution of the first bounce to bounce(π) is $n - \alpha_1 = \alpha_2 + \ldots + \alpha_b$, the contribution of the second bounce is $n - \alpha_1 - \alpha_2 = \alpha_3 + \ldots + \alpha_b$, et cetera, so bounce($\pi$) $= \alpha_2 + 2\alpha_3 + \ldots + (b-1)\alpha_b$. $\square$

Plethystic Formulas for the q, t-Catalan

The recurrence for the $F_{n,k}(q, t)$ led Garsia and Haglund to search for a corresponding recurrence involving the ∇ operator, which eventually resulted in the following.

THEOREM 3.5. *For any integers $k, m \geq 1$,*

(3.11)
$$\left\langle \nabla e_m \left[X \frac{1 - q^k}{1 - q} \right], e_m \right\rangle = \sum_{r=1}^{m} \begin{bmatrix} r + k - 1 \\ r \end{bmatrix} t^{m-r} q^{\binom{r}{2}} \left\langle \nabla e_{m-r} \left[X \frac{1 - q^r}{1 - q} \right], e_{m-r} \right\rangle.$$

We will give a detailed discussion of the proof of a generalization of Theorem 3.5 in Chapter 6. For now we list some of its consequences.

COROLLARY 3.5.1. *For all $0 \leq k \leq n$,*

(3.12)
$$F_{n,k}(q, t) = t^{n-k} q^{\binom{k}{2}} \left\langle \nabla e_{n-k} \left[X \frac{1 - q^k}{1 - q} \right], e_{n-k} \right\rangle.$$

PROOF. Let

(3.13)
$$Q_{n,k}(q, t) = t^{n-k} q^{\binom{k}{2}} \left\langle \nabla e_{n-k} \left[X \frac{1 - q^k}{1 - q} \right], e_{n-k} \right\rangle.$$

One easily checks that Theorem 3.5 implies

(3.14)
$$Q_{n,k}(q, t) = t^{n-k} q^{\binom{k}{2}} \sum_{r=0}^{n-k} \begin{bmatrix} r + k - 1 \\ r \end{bmatrix} Q_{n-k,r}(q, t), \qquad Q_{n,0} = \chi(n = 0).$$

Thus $Q_{n,k}(q, t)$ satisfies the same recurrence and initial conditions as $F_{n,k}(q, t)$, and hence the two are equal. $\square$

Theorem 3.2 is a special case of Corollary 3.5.1. To see why, note that a path β in $L_{n+1,n+1}^+(1)$ can be identified with a path π in $L_{n,n}^+$ by removing the first N and E steps of β. Clearly area(β) = area(π), and bounce(β) = bounce(π) + n. Thus $F_{n+1,1}(q, t) = t^n F_n(q, t)$. By Corollary 3.5.1 we have

(3.15)
$$t^n F_n(q, t) = F_{n+1,1}(q, t) = t^n \left\langle \nabla e_n \left[X \frac{1 - q}{1 - q} \right], e_n \right\rangle = t^n \left\langle \nabla e_n, e_n \right\rangle$$

and Theorem 3.2 follows upon comparing the opposite ends of (3.15).

Using the following result we can express $F_{n,k}(q, t)$ as an explicit sum of rational functions [GH02].

PROPOSITION 3.5.1.

(3.16)
$$\tilde{H}_\mu[(1 - t)(1 - q^k); q, t] = \Pi_\mu h_k[(1 - t)B_\mu](1 - q^k).$$

PROOF. Letting $\lambda = (k)$ in (2.61) gives

$$(3.17) \qquad \frac{\tilde{H}_\mu[MB_{(k)}; q, t]}{\Pi_\mu} = \frac{\tilde{H}_{(k)}[MB_\mu; q, t]}{\Pi_{(k)}}.$$

Since $B_{(k)} = (1 - q^k)/(1 - q)$ and $\Pi_{(k)} = (q)_{k-1}$, using (2.43) we see (3.17) reduces to (3.16). $\qquad\square$

COROLLARY 3.5.2. *For* $n \geq 1$, $k \geq 0$,

$$(3.18) \qquad e_n\left[X\frac{1 - q^k}{1 - q}\right] = \sum_{\mu \vdash n} \frac{(1 - q^k)\tilde{H}_\mu h_k[(1 - t)B_\mu]\Pi_\mu}{w_\mu}.$$

PROOF. Letting $Y = (1 - t)(1 - q^k)$ in (2.69) we get

$$(3.19) \qquad e_n\left[X\frac{1 - q^k}{1 - q}\right] = \sum_{\mu \vdash m} \frac{\tilde{H}_\mu[X; q, t]\tilde{H}_\mu[(1 - t)(1 - q^k); q, t]}{w_\mu}.$$

Eq. (3.18) now follows from (3.16). $\qquad\square$

THEOREM 3.6. *For* $1 \leq k \leq n$,

$$(3.20) \qquad F_{n,k}(q, t) = t^{n-k}q^{\binom{k}{2}} \sum_{\mu \vdash n-k} \frac{(1 - q^k)T_\mu^2 h_k[(1 - t)B_\mu]\Pi_\mu}{w_\mu}.$$

PROOF. Using (3.12) we have

$$(3.21) \qquad F_{n,k}(q, t) = t^{n-k}q^{\binom{k}{2}} \left\langle \nabla e_{n-k}[X\frac{1 - q^k}{1 - q}], e_{n-k} \right\rangle$$

$$(3.22) \qquad\qquad = t^{n-k}q^{\binom{k}{2}} \sum_{\mu \vdash n-k} \frac{(1 - q^k)T_\mu \left\langle \tilde{H}_\mu, e_{n-k} \right\rangle h_k[(1 - t)B_\mu]\Pi_\mu}{w_\mu}$$

$$(3.23) \qquad\qquad = t^{n-k}q^{\binom{k}{2}} \sum_{\mu \vdash n-k} \frac{(1 - q^k)T_\mu^2 h_k[(1 - t)B_\mu]\Pi_\mu}{w_\mu},$$

by (3.18) and the $d = 0$ case of (2.44). $\qquad\square$

There is an alternate way of expressing Corollary 3.5.1 involving what could be called "q-Taylor coefficients". Define symmetric functions $E_{n,k} = E_{n,k}(X; q, t)$ by means of the equation

$$(3.24) \qquad e_n\left[X\frac{1 - z}{1 - q}\right] = \sum_{k=1}^{n} \frac{(z; q)_k}{(q)_k} E_{n,k}, \qquad n \geq 1.$$

PROPOSITION 3.6.1.

$$(3.25) \qquad \langle \nabla E_{n,k}, e_n \rangle = t^{n-k}q^{\binom{k}{2}} \left\langle \nabla e_{n-k}\left[X\frac{1 - q^k}{1 - q}\right], e_{n-k} \right\rangle$$

$$(3.26) \qquad\qquad = F_{n,k}(q, t).$$

PROOF. Letting $z = q^p$ and $n = m$ in (3.24) we get

$$(3.27) \qquad \left\langle \nabla e_m\left[X\frac{1 - q^p}{1 - q}\right], e_m \right\rangle = \sum_{r=1}^{m} \frac{(q^p)_r}{(q)_r} \langle \nabla E_{m,r}, e_m \rangle.$$

On the other hand, from Theorem 3.5 we have

(3.28)
$$\left\langle \nabla e_m \left[X \frac{1-q^p}{1-q} \right], e_m \right\rangle$$
$$= \sum_{r=1}^{m} \begin{bmatrix} r+p-1 \\ r \end{bmatrix} t^{m-r} q^{\binom{r}{2}} \left\langle \nabla e_{m-r} \left[X \frac{1-q^r}{1-q} \right], e_{m-r} \right\rangle.$$

Thus

(3.29)
$$\sum_{r=1}^{m} \frac{(q^p)_r}{(q)_r} \left\langle \nabla E_{m,r}, e_m \right\rangle$$
$$= \sum_{r=1}^{m} \begin{bmatrix} r+p-1 \\ r \end{bmatrix} t^{m-r} q^{\binom{r}{2}} \left\langle \nabla e_{m-r} \left[X \frac{1-q^r}{1-q} \right], e_{m-r} \right\rangle.$$

Since $\begin{bmatrix} r+p-1 \\ r \end{bmatrix} = \frac{(q^p)_r}{(q)_r}$, (3.29) can be viewed as a polynomial identity between the two sides of (3.29), in the variable $w = q^p$. They agree for all $p \in \mathbb{N}$, which gives infinitely many common values to the polynomials and forces them to be equal. Since $(w)_r$ is a polynomial of degree r, the $\{(w)_r, r \geq 1\}$ are linearly independent, so the coefficients of $(w)_r$ on both sides of (3.29) must agree. $\qquad\square$

REMARK 3.7. In the proof of Proposition 3.25 we replace z by $z = q^p$ in (3.24), an equation involving plethystic brackets. This can be justified by writing

(3.30)
$$e_n \left[X \frac{1-z}{1-q} \right] = \sum_{\lambda \vdash n} c_\lambda \prod_i p_{\lambda_i}(X) \frac{1-z^{\lambda_i}}{1-q^{\lambda_i}}$$

for some c_λ independent of X. It is now clear that z can be replaced by any power of a real parameter on both sides of (3.24). We could not, however, replace z by say $q + t$ in (3.24), since if we did the left-hand-side would become

(3.31)
$$\sum_{\lambda \vdash n} c_\lambda \prod_i p_{\lambda_i}(X) \frac{1-q^{\lambda_i}-t^{\lambda_i}}{1-q^{\lambda_i}} \neq \sum_{\lambda \vdash n} c_\lambda \prod_i p_{\lambda_i}(X) \frac{1-(q+t)^{\lambda_i}}{1-q^{\lambda_i}}.$$

Decomposing e_n into the $E_{n,k}$ also has applications to the study of $\langle \nabla e_n, s_\lambda \rangle$ for general λ. We mention in particular the following conjecture, which A. Garsia and the author have tested in Maple. Refinements and generalizations of this conjecture will be discussed in Chapters 5 and 6.

CONJECTURE 3.8. For $1 \leq k \leq n$ and $\lambda \in Par(n)$,

(3.32)
$$\nabla E_{n,k} \in \mathbb{N}[q,t].$$

EXERCISE 3.9.   (1) By letting z be an appropriate power of q in (3.24), show that

(3.33)
$$E_{n,1} = \frac{1}{(-q)^{n-1}} h_n[X].$$

(2) By letting $E = X/(1-q)$ and $F = 1-z$ in (1.67), and then using (1.72), show that

(3.34)
$$\nabla E_{n,n} = (q)_n h_n \left[\frac{X}{1-q} \right] = \tilde{H}_n[X; q, t].$$

Thus $\nabla E_{n,n}$ is the character for the space H_n of harmonics in one set of variables.

At the end of Chapter 7 we outline a proof of the following result from [**Hag04b**], which shows Conjecture 3.8 holds for $k = n - 1$.

PROPOSITION 3.9.1. *For any* $\lambda \in Par(n)$,

$$(3.35) \qquad \langle \nabla E_{n,n-1}, s_\lambda \rangle = t \begin{bmatrix} n-1 \\ 1 \end{bmatrix} \sum_{\substack{T \in SYT(\lambda) \\ n \text{ is above } n-1 \text{ in } T}} q^{maj(T)-n+1}.$$

For example, $\langle \nabla E_{4,3}, s_{211} \rangle$ would involve the sum over $SYT(211)$ in which 4 is in a row above 3. There are two such tableaux, namely

$$(3.36) \qquad \begin{matrix} 4 & & & 4 & \\ 3 & & & 2 & \\ 1 & 2 & & 1 & 3 \end{matrix}$$

The maj of the tableau on the left is $2 + 3 = 5$, and of the one on the right is $1 + 3 = 4$. Thus we get

$$(3.37) \qquad \langle \nabla E_{4,3}, s_{211} \rangle = t(1 + q + q^2)(q^2 + q).$$

The Special Values $t = 1$ and $t = 1/q$

Garsia and Haiman proved that

$$(3.38) \qquad C_n(q, 1) = C_n(q)$$

$$(3.39) \qquad q^{\binom{n}{2}} C_n(q, 1/q) = \frac{1}{[n+1]} \begin{bmatrix} 2n \\ n \end{bmatrix},$$

which shows that both the Carlitz-Riordan and MacMahon q-Catalan numbers are special cases of $C_n(q, t)$. In this section we derive analogous results for $F_{n,k}(q, 1)$ and $F_{n,k}(q, 1/q)$.

By definition we have

$$(3.40) \qquad F_n(q, 1) = C_n(q).$$

It is perhaps worth mentioning that the $F_{n,k}(q, 1)$ satisfy the simple recurrence

$$(3.41) \qquad F_{n,k}(q, 1) = \sum_{m=k}^{n} q^{m-1} F_{m-1,k-1}(q, 1) F_{n-m}(q, 1).$$

This follows by grouping paths in $L_{n,n}^+(k)$ according to the first time they return to the diagonal, at say (m, m), then arguing as in the proof of Proposition 1.20.

The $F_{n,k}(q, 1/q)$ satisfy the following refinement of (3.39).

THEOREM 3.10. *For* $1 \leq k \leq n$,

$$(3.42) \qquad q^{\binom{n}{2}} F_{n,k}(q, 1/q) = \frac{[k]}{[n]} \begin{bmatrix} 2n-k-1 \\ n-k \end{bmatrix} q^{(k-1)n}.$$

PROOF. Since $F_{n,n}(q,t) = q^{\binom{n}{2}}$, Theorem 3.10 holds for $k = n$. If $1 \leq k < n$, we start with Theorem 3.4 and then use induction on n;

(3.43)

$$q^{\binom{n}{2}}F_{n,k}(q,q^{-1}) = q^{\binom{n}{2}}q^{-\binom{n-k}{2}}\sum_{r=1}^{n-k}q^{\binom{n-k}{2}}F_{n-k,r}(q,q^{-1})q^{\binom{k}{2}-(n-k)}\begin{bmatrix}r+k-1\\r\end{bmatrix}$$

(3.44)

$$= q^{\binom{n}{2}+\binom{k}{2}-(n-k)}q^{-\binom{n-k}{2}}\sum_{r=1}^{n-k}\begin{bmatrix}r+k-1\\r\end{bmatrix}\frac{[r]}{[n-k]}\begin{bmatrix}2(n-k)-r-1\\n-k-r\end{bmatrix}q^{(r-1)(n-k)}$$

(3.45)

$$= q^{(k-1)n}\sum_{r=1}^{n-k}\begin{bmatrix}r+k-1\\r\end{bmatrix}\frac{[r]}{[n-k]}\begin{bmatrix}2(n-k)-2-(r-1)\\n-k-1-(r-1)\end{bmatrix}q^{(r-1)(n-k)}$$

(3.46)

$$= q^{(k-1)n}\frac{[k]}{[n-k]}\sum_{u=0}^{n-k-1}\begin{bmatrix}k+u\\u\end{bmatrix}q^{u(n-k)}\begin{bmatrix}2(n-k)-2-u\\n-k-1-u\end{bmatrix}.$$

Using (1.28) we can write the right-hand side of (3.46) as

(3.47)

$$q^{(k-1)n}\frac{[k]}{[n-k]}\frac{1}{(zq^{n-k})_{k+1}}\frac{1}{(z)_{n-k}}\Big|_{z^{n-k-1}} = q^{(k-1)n}\frac{[k]}{[n-k]}\frac{1}{(z)_{n+1}}\Big|_{z^{n-k-1}}$$

(3.48)

$$= q^{(k-1)n}\frac{[k]}{[n-k]}\begin{bmatrix}n+n-k-1\\n\end{bmatrix}.$$

$\square$

COROLLARY 3.10.1.

(3.49)
$$q^{\binom{n}{2}}F_n(q,1/q) = \frac{1}{[n+1]}\begin{bmatrix}2n\\n\end{bmatrix}.$$

PROOF. N. Loehr has pointed out that we can use

(3.50)
$$F_{n+1,1}(q,t) = t^n F_n(q,t),$$

which by Theorem 3.10 implies

(3.51)
$$q^{\binom{n+1}{2}}F_{n+1,1}(q,1/q) = \frac{[1]}{[n+1]}\begin{bmatrix}2(n+1)-2\\n+1-1\end{bmatrix} = \frac{[1]}{[n+1]}\begin{bmatrix}2n\\n\end{bmatrix}$$

$$= q^{\binom{n+1}{2}-n}F_n(q,1/q) = q^{\binom{n}{2}}F_n(q,1/q).$$

$\square$

The Symmetry Problem and the dinv Statistic

From its definition, it is easy to show $C_n(q,t) = C_n(t,q)$, since the arm and leg values for μ equal the leg and arm values for μ', respectively, which implies q and t are interchanged when comparing terms in (2.53) corresponding to μ and μ'. This also follows from the theorem that $C_n(q,t) = \mathcal{H}(DH_n^\epsilon; q, t)$. Thus we have

(3.52)
$$\sum_{\pi\in L_{n,n}^+}q^{\text{area}(\pi)}t^{\text{bounce}(\pi)} = \sum_{\pi\in L_{n,n}^+}q^{\text{bounce}(\pi)}t^{\text{area}(\pi)},$$

a surprising statement in view of the apparent dissimilarity of the area and bounce statistics. At present there is no other known way to prove (3.52) other than as a corollary of Theorem 3.2.

OPEN PROBLEM 3.11. *Prove (3.52) by exhibiting a bijection on Dyck paths which interchanges area and bounce.*

A solution to Problem 3.11 should lead to a deeper understanding of the combinatorics of DH_n. We now give a bijective proof from [**Hag03**] of a very special case of (3.52), by showing that the marginal distributions of area and bounce are the same, i.e. $F_n(q,1) = F_n(1,q)$.

THEOREM 3.12.

$$(3.53) \qquad \sum_{\pi \in L_{n,n}^+} q^{area(\pi)} = \sum_{\pi \in L_{n,n}^+} q^{bounce(\pi)}.$$

PROOF. Given $\pi \in L_{n,n}^+$, let $a_1 a_2 \cdots a_n$ denote the sequence whose ith element is the ith coordinate of the area vector of π, i.e. the length of the ith row (from the bottom) of π. A moment's thought shows that such a sequence is characterized by the property that it begins with zero, consists of n nonnegative integers, and has no 2-ascents, i.e. values of i for which $a_{i+1} > a_i + 1$. To construct such a sequence we begin with an arbitrary multiset of row lengths, say $\{0^{\alpha_1} 1^{\alpha_2} \cdots (b-1)^{\alpha_b}\}$ and then choose a multiset permutation τ of $\{0^{\alpha_1} 1^{\alpha_2}\}$ which begins with 0 in $\binom{\alpha_1 - 1 + \alpha_2}{\alpha_2}$ ways. Next we will insert the α_3 twos into τ, the requirement of having no 2-ascents translating into having no consecutive 02 pairs. This means the number of ways to do this is $\binom{\alpha_2 - 1 + \alpha_3}{\alpha_3}$, independent of the choice of τ. The formula

$$(3.54) \qquad F_n(q,1) = \sum_{b=1}^{n} \sum_{\substack{\alpha_1 + \ldots + \alpha_b = n \\ \alpha_i > 0}} q^{\sum_{i=2}^{b} \alpha_i (i-1)} \prod_{i=1}^{b-1} \binom{\alpha_i - 1 + \alpha_{i+1}}{\alpha_{i+1}}$$

follows, since the product above counts the number of Dyck paths with a specified multiset of row lengths, and the power of q is the common value of area for all these paths. Comparing (3.54) with the $q = 1, t = q$ case of (3.10) completes the proof. □

EXERCISE 3.13. Prove combinatorially that

$$(3.55) \qquad F_n(q,t)|_t = F_n(t,q)|_t.$$

There is another pair of statistics for the q, t-Catalan discovered by M. Haiman [**Hai00a**]. It involves pairing area with a different statistic we call dinv, for "diagonal inversion" or "d-inversion". It is defined, with a_i the length of the ith row from the bottom, as follows.

DEFINITION 3.14. Let $\pi \in L_{n,n}^+$. Let

$$\text{dinv}(\pi) = |\{(i,j) : 1 \le i < j \le n \quad a_i = a_j\}|$$
$$+ |\{(i,j) : 1 \le i < j \le n \quad a_i = a_j + 1\}|.$$

In words, $\text{dinv}(\pi)$ is the number of pairs of rows of π of the same length, or which differ by one in length, with the longer row below the shorter. For example, for the path on the left in Figure 3, with row lengths on the right, the inversion pairs (i,j) are $(2,7), (3,4), (3,5), (3,8), (4,5), (4,8), (5,8)$ (corresponding to pairs of rows of the same length) and $(3, 7), (4, 7), (5, 7), (6, 8)$ (corresponding to rows which differ by one in length), thus dinv $= 11$. We call inversion pairs between rows of the same length "equal-length" inversions, and the other kind "offset-length" inversions.

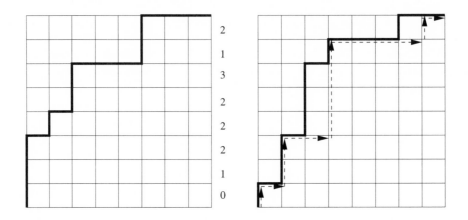

FIGURE 3. A path π with row lengths to the right, and the image $\zeta(\pi)$.

THEOREM 3.15.

$$(3.56) \qquad \sum_{\pi \in L_{n,n}^+} q^{dinv(\pi)} t^{area(\pi)} = \sum_{\pi \in L_{n,n}^+} q^{area(\pi)} t^{bounce(\pi)}.$$

PROOF. We will describe a bijective map ζ on Dyck paths with the property that

$$(3.57) \qquad\qquad dinv(\pi) = area(\zeta(\pi))$$
$$area(\pi) = bounce(\zeta(\pi)).$$

Say $b-1$ is the length of the longest row of π. The lengths of the bounce steps of ζ will be $\alpha_1, \ldots, \alpha_b$, where α_i is the number of rows of length $i-1$ in π. To construct the actual path ζ, place a pen at the lattice point $(\alpha_1, \alpha_1 + \alpha_2)$ (the second peak of the bounce path of ζ). Start at the end of the area sequence and travel left. Whenever you encounter a 1, trace a South step with your pen. Whenever you encounter a 0, trace a West step. Skip over all other numbers. Your pen will end up at the top of the first peak of ζ. Now go back to the end of the area sequence, and place your pen at the top of the third peak. Traverse the area sequence again from right to left, but this time whenever you encounter a 2 trace out a South step, and whenever you encounter a 1, trace out a West step. Skip over any other numbers. Your pen will end up at the top of the second peak of ζ. Continue at the top of the fourth peak looking at how the rows of length 3 and 2 are interleaved, etc.

It is easy to see this map is a bijection, since given ζ, from the bounce path we can determine the multiset of row lengths of π. We can then build up the area sequence of π just as in the proof of Theorem 3.12. From the portion of the path between the first and second peaks we can see how to interleave the rows of lengths 0 and 1, and then we can insert the rows of length 2 into the area sequence, etc.

Note that when tracing out the part of ζ between the first and second peaks, whenever we encounter a 0 and trace out a West step, the number of area squares directly below this West step and above the Bounce path of ζ equals the number of 1's to the left of this 0 in the area sequence, which is the number of offset-length inversion pairs involving the corresponding row of length 0. Since the area below the bounce path clearly counts the total number of equal-length inversions, it follows that $\operatorname{dinv}(\pi) = \operatorname{area}(\zeta(\pi))$.

Now by direct calculation,

$$(3.58) \qquad \operatorname{bounce}(\zeta) = n - \alpha_1 + n - \alpha_1 - \alpha_2 + \ldots n - \alpha_1 - \ldots - \alpha_{b-1}$$

$$= (\alpha_2 + \ldots + \alpha_b) + \ldots + (\alpha_b) = \sum_{i=1}^{b-1} i\alpha_{i+1} = \operatorname{area}(\pi).$$

$\square$

REMARK 3.16. The construction of the bounce path for a Dyck path occurs in an independent context, in work of Andrews, Krattenthaler, Orsina and Papi [**AKOP02**] on the enumeration of ad-nilpotent ideals of a Borel subalgebra of $\operatorname{sl}(n+1, \mathbb{C})$. They prove the number of times a given nilpotent ideal needs to be bracketed with itself to become zero equals the number of bounces of the bounce path of a certain Dyck path associated to the ideal. Another of their results is a bijective map on Dyck paths which sends a path with b bounces to a path whose longest row is of length $b - 1$. The ζ map above is just the inverse of their map. Because they only considered the number of bounces, and not the bounce statistic per say, they did not notice any connection between $C_n(q, t)$ and their construction.

Theorem 3.2 now implies

COROLLARY 3.16.1.

$$(3.59) \qquad C_n(q, t) = \sum_{\pi \in L_{n,n}^+} q^{dinv(\pi)} t^{area(\pi)}.$$

We also have

$$(3.60) \qquad F_{n,k}(q, t) = \sum_{\substack{\pi \in L_{n,n}^+ \\ \pi \text{ has exactly } k \text{ rows of length } 0}} q^{dinv(\pi)} t^{area(\pi)},$$

since under the ζ map, paths with k rows of length 0 correspond to paths whose first bounce step is of length k.

REMARK 3.17. N. Loehr had noted that if one could find a map which fixes area and interchanges dinv and bounce, by combining this with the ζ map one would have a map which interchanges area and bounce, solving Problem 3.11. S. Mason [**Mas06**] has provided the following example to show this is impossible. Figure 4 contains the 5 paths for $n = 6$ with area $= 11$, with row lengths on the right. By inspection the (dinv, bounce) pairs are not symmetric.

EXERCISE 3.18. Let

$$(3.61) \qquad G_{n,k}(q, t) = \sum_{\substack{\pi \in L_{n,n}^+ \\ \pi \text{ has exactly } k \text{ rows of length } 0}} q^{\operatorname{dinv}(\pi)} t^{\operatorname{area}(\pi)}.$$

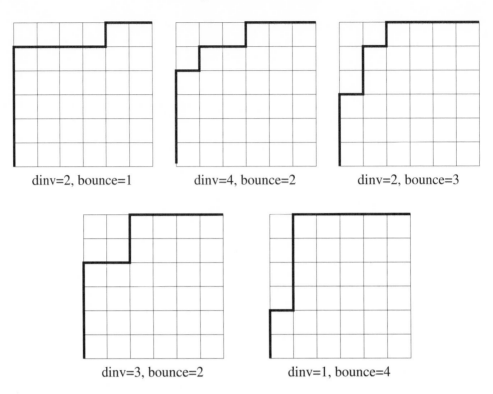

FIGURE 4. Paths for $n = 6$ with area $= 11$ and their dinv and bounce values.

Without referencing any results on the bounce statistic, prove combinatorially that

$$(3.62) \qquad G_{n,k}(q,t) = q^{n-k}t^{\binom{k}{2}} \sum_{r=1}^{n-k} \begin{bmatrix} r + k - 1 \\ r \end{bmatrix}_t G_{n-k,r}(q,t).$$

EXERCISE 3.19. Haiman's conjectured statistics for $C_n(q,t)$ actually involved a different description of dinv. Let $\lambda(\pi)$ denote the partition consisting of the $\binom{n}{2} - \text{area}(\pi)$ squares above π but inside the $n \times n$ square. (This is the Ferrers graph of a partition in the so-called English convention, which is obtained from the French convention of Figure 1 by reflecting the graph about the x-axis. In this convention, the leg of a square s is defined as the number of squares of λ below s in the column and above the lower border π, and the arm as the number of squares of λ to the right and in the row.) Then Haiman's original version of dinv was the number of cells s of λ for which

$$(3.63) \qquad \text{leg}(s) \le \text{arm}(s) \le \text{leg}(s) + 1.$$

Prove this definition of dinv is equivalent to Definition 3.14.

q-Lagrange Inversion

When working with ∇e_n, q-Lagrange inversion is useful in dealing with the special case $t = 1$. In this section we derive a general q-Lagrange inversion theorem based on work of Garsia and Haiman. We will be working in the ring of formal power series, and we begin with a result of Garsia [**Gar81**].

THEOREM 3.20. *If*

$$(3.64) \qquad (F \circ_q G)(z) = \sum_n f_n G(z) G(qz) \cdots G(q^{n-1}z),$$

where $F = \sum_n f_n z^n$, then

$$(3.65) \qquad F \circ_q G = z \ and \ G \circ_{q^{-1}} F = z$$

are equivalent to each other and also to

$$(3.66) \qquad (\Phi \circ_{q^{-1}} F) \circ_q G = \Phi = (\Phi \circ_q G) \circ_{q^{-1}} F \qquad for \ all \ \Phi.$$

Given $\pi \in L_{n,n}^+$, let $\beta(\pi) = \beta_1(\pi)\beta_2(\pi)\cdots$ denote the partition consisting of the vertical step lengths of π (i.e. the lengths of the maximal blocks of consecutive 0's in $\sigma(\pi)$), arranged in nonincreasing order. For example, for the path on the left in Figure 3 we have $\beta = (3,2,2,1)$. By convention we set $\beta(\emptyset) = \emptyset$. Define $H(z)$ via the equation $1/H(-z) := \sum_{k=0}^{\infty} e_k z^k$. Using Theorem 3.20, Haiman [**Hai94**, pp. 47-48] derived the following.

THEOREM 3.21. *Define $h_n^*(q)$, $n \geq 0$ via the equation*

$$(3.67) \qquad \sum_{k=0}^{\infty} e_k z^k = \sum_{n=0}^{\infty} q^{-\binom{n}{2}} h_n^*(q) z^n H(-q^{-1}z) H(-q^{-2}z) \cdots H(-q^{-n}z).$$

Then for $n \geq 0$, $h_n^*(q)$ has the explicit expression

$$(3.68) \qquad h_n^*(q) = \sum_{\pi \in L_{n,n}^+} q^{area(\pi)} e_{\beta(\pi)}.$$

For example, we have

$$(3.69) \qquad h_3^*(q) = q^3 e_3 + q^2 e_{2,1} + 2q e_{2,1} + e_{1^3}.$$

We now derive a slight generalization of Theorem 3.21 which stratifies Dyck paths according to the length of their first bounce step.

THEOREM 3.22. *Let c_k, $k \geq 0$ be a set of variables. Define $h_n^*(\mathbf{c}, q)$, $n \geq 0$ via the equation*

$$(3.70) \qquad \sum_{k=0}^{\infty} e_k c_k z^k = \sum_{n=0}^{\infty} q^{-\binom{n}{2}} h_n^*(\mathbf{c}, q) z^n H(-q^{-1}z) H(-q^{-2}z) \cdots H(-q^{-n}z).$$

Then for $n \geq 0$, $h_n^*(\mathbf{c}, q)$ has the explicit expression

$$(3.71) \qquad h_n^*(\mathbf{c}, q) = \sum_{k=0}^{n} c_k \sum_{\pi \in L_{n,n}^+(k)} q^{area(\pi)} e_{\beta(\pi)}.$$

For example, we have

$$(3.72) \qquad h_3^*(\mathbf{c}, q) = q^3 e_3 c_3 + (q^2 e_{2,1} + q e_{2,1}) c_2 + (q e_{2,1} + e_{1^3}) c_1.$$

Proof. Our proof follows Haiman's proof of Theorem 3.21 closely. Set $H^*(z,\mathbf{c};q) := \sum_{n=0}^{\infty} h_n^*(\mathbf{c},q)z^n$, $H^*(z;q) := \sum_{n=0}^{\infty} h_n^*(q)z^n$, $\Phi = H^*(zq,\mathbf{c};q)$, $F = zH(-z)$, and $G = zH^*(qz;q)$. Replacing z by zq in (3.70) we see Theorem 3.22 is equivalent to the statement

$$(3.73) \quad \sum_{k=0}^{\infty} e_k c_k q^k z^k = \sum_{n=0}^{\infty} q^n h_n^*(\mathbf{c},q)zH(-z)zq^{-1}H(-q^{-1}z)\cdots zq^{1-n}H(-q^{1-n}z)$$

$$= \Phi \circ_{q^{-1}} F.$$

On the other hand, Theorem 3.21 can be expressed as

$$(3.74) \quad \frac{1}{H(-z)} = \sum_{n=0}^{\infty} q^{-\binom{n}{2}} h_n^*(q)z^n H(-z/q)\cdots H(-z/q^n),$$

or

$$(3.75) \quad z = \sum_{n=0}^{\infty} q^{-\binom{n}{2}} h_n^*(q)z^{n+1} H(-z)H(-z/q)\cdots H(-z/q^n)$$

$$= \sum_{n=0}^{\infty} q^n h_n^*(q)\{zH(-z)\}\left\{\frac{z}{q}H(-z/q)\right\}\cdots\left\{\frac{z}{q^n}H(-z/q^n)\right\}$$

$$(3.76) \quad = G \circ_{q^{-1}} F.$$

Thus, using Theorem 3.20, we have

$$(3.77) \quad \Phi = (\Phi \circ_{q^{-1}} F) \circ_q G$$

$$= \left(\sum_{k=0}^{\infty} e_k \mu_k q^k z^k\right) \circ_q G.$$

Comparing coefficients of z^n in (3.77) and simplifying we see that Theorem 3.22 is equivalent to the statement

$$(3.78) \quad q^n h_n^*(\mathbf{c},q) = \sum_{k=0}^{n} q^{\binom{k}{2}+k} e_k c_k \sum_{\substack{n_1+\ldots+n_k=n-k \\ n_i \geq 0}} q^{n-k} \prod_{i=1}^{k} q^{(i-1)n_i} h_{n_i}^*(q).$$

To prove (3.78) we use the "factorization of Dyck paths" as discussed in [Hai94]. This can be be represented pictorially as in Figure 5. The terms multiplied by c_k correspond to $\pi \in L_{n,n}^+(k)$. The area of the trapezoidal region whose left border is on the line $y = x + i$ is $(i-1)n_i$, where n_i is the length of the left border of the trapezoid. Using the fact that the path to the left of this trapezoid is in L_{n_i,n_i}^+, (3.78) now becomes transparent. □

Letting $e_k = 1$, $c_j = \chi(j=k)$ and replacing q by q^{-1} and z by z/q in Theorem 3.22 we get the following.

Corollary 3.22.1. For $1 \leq k \leq n$,

$$(3.79) \quad z^k = \sum_{n \geq k} q^{\binom{n}{2}-n+k} F_{n,k}(q^{-1},1)z^n (z)_n.$$

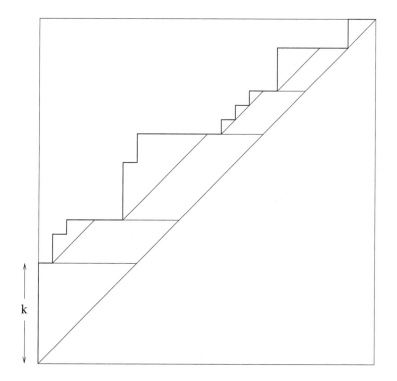

FIGURE 5. A Dyck path factored into smaller paths.

Theorem 3.22 is a *q*-analogue of the general Lagrange inversion formula [**AAR99**, p.629]

$$(3.80) \qquad f(x) = f(0) + \sum_{n=1}^{\infty} \frac{x^n}{n!\phi(x)^n} \left[\frac{d^{n-1}}{dx^{n-1}} (f'(x)\phi^n(x)) \right]_{x=0},$$

where ϕ and f are analytic in a neighborhood of 0, with $\phi(0) \neq 0$. To see why, assume WLOG $f(0) = 1$, and set $f(x) = \sum_{k=0}^{\infty} c_k x^k$ and $\phi = \frac{1}{H(-x)} = \sum_{k=0}^{\infty} e_k x^k$ in (3.80) to get

$$(3.81) \quad \sum_{k=1}^{\infty} c_k x^k = \sum_{n=1}^{\infty} \frac{x^n}{n!} H(-x)^n \frac{d^{n-1}}{dx^{n-1}} \left(\sum_{k=1}^{\infty} kc_k x^{k-1} (\sum_{m=0}^{\infty} e_m x^m)^n \right) |_{x=0}$$

$$(3.82) \qquad = \sum_{n=1}^{\infty} \frac{x^n}{n!} H(-x)^n \sum_{k=1}^{n} kc_k (n-1)! (\sum_{m=0}^{\infty} e_m x^m)^n |_{x^{n-k}}$$

$$(3.83) \qquad = \sum_{n=1}^{\infty} x^n H(-x)^n \sum_{k=1}^{n} c_k \frac{k}{n} \sum_{\substack{j_1+j_2+\ldots+j_n=n-k \\ j_i \geq 0}} e_{j_1} e_{j_2} \cdots e_{j_n}$$

$$(3.84) \qquad = \sum_{n=1}^{\infty} x^n H(-x)^n \sum_{k=1}^{n} c_k \frac{k}{n} \sum_{\alpha \vdash n-k} e_\alpha \binom{n}{k-\ell(\alpha), n_1(\alpha), n_2(\alpha), \ldots}.$$

The equivalence of (3.84) to the $q = 1$ case of Theorem 3.22 (with c_k replaced by c_k/e_k) will follow if we can show that for any fixed $\alpha \vdash n - k$,

$$(3.85) \qquad \sum_{\substack{\pi \in L_{n,n}^+(k) \\ \beta(\pi) - k = \alpha}} 1 = \frac{k}{n} \binom{n}{k - \ell(\alpha), n_1(\alpha), n_2(\alpha), \ldots},$$

where $\beta - k$ is the partition obtained by removing one part of size k from β. See [Hag03] for an inductive proof of (3.85).

In [Hai94] Haiman includes a discussion of the connection of Theorem 3.21 to q-Lagrange inversion formulas of Andrews, Garsia, and Gessel [And75], [Gar81], [Ges80]. Further background on these formulas is contained in [Sta88]. We should also mention that in [GH96a] Garsia and Haiman derive the formula

$$(3.86) \qquad h_n^*(q) = \sum_{\mu \vdash n} \left(\prod_i q^{\binom{\mu_i}{2}} h_{\mu_i} \left[\frac{X}{1 - q} \right] \right) f_\mu [1 - q],$$

where $f_\mu[X]$ is the so-called "forgotten" basis of symmetric functions, defined by

$$(3.87) \qquad \omega m_\beta = f_\beta.$$

They then use the fact that

$$(3.88) \qquad \tilde{H}_\mu[X; q, 1] = \prod_{i=1}^{\ell(\mu)} q^{\binom{\mu_i}{2}} h_{\mu_i} \left[\frac{X}{1 - q} \right]$$

to show that the limit as $t \to 1^-$ of ∇e_n also equals $h_n^*(q)$. This gives the interesting identity

$$(3.89) \qquad \mathcal{F}(DH_n; q, 1) = \sum_{\pi \in L_{n,n}^+} q^{\text{area}(\pi)} e_{\beta(\pi)}$$

for the $t = 1$ case of the Frobenius series. The product $e_{\beta(\pi)}$ above can be expanded in terms of Schur functions by iterating the Pieri rule (Theorem 1.20 (4)) in which case the coefficient of s_λ is simply $K_{\lambda', \beta}$. One way to view the fundamental problem 2.15 is to find a way to associate appropriate powers of t, reflecting in some way the overall shape of π, to each of the elements of $SSYT(\lambda', \beta(\pi))$.

Let $^q\nabla$ be the linear operator on symmetric functions defined on the Schur function basis by $^q\nabla s_\lambda = \nabla s_\lambda|_{t=1}$. In [BGHT99] it is shown that for any two symmetric functions P, Q,

$$(3.90) \qquad {}^q\nabla(PQ) = ({}^q\nabla P)({}^q\nabla Q).$$

Using plethystic manipulations, in [GH02] the following is derived.

$$(3.91) \qquad {}^q\nabla e_m \left[X \frac{1 - q^k}{1 - q} \right] = \sum_{r=0}^m e_r \left[X \frac{1 - q^k}{1 - q} \right] q^{\binom{r}{2}} \, {}^q\nabla e_{m-r} \left[X \frac{1 - q^r}{1 - q} \right],$$

and moreover

$$(3.92) \qquad e_k q^{\binom{k}{2}} \, {}^q\nabla e_{n-k} \left[X \frac{1 - q^k}{1 - q} \right] = \sum_{\pi \in L_{n,n}^+(k)} q^{\text{area}(\pi)} e_{\beta(\pi)}.$$

This implies Theorem 3.22 can be rephrased as follows.

THEOREM 3.23. *Let $c_k, k \geq 0$ be a set of variables. Then*

(3.93)
$$\sum_{k=0}^{\infty} e_k c_k z^k =$$

$$\sum_{n=0}^{\infty} q^{-\binom{n}{2}} z^n H(-q^{-1}z) H(-q^{-2}z) \cdots H(-q^{-n}z) \sum_{k=0}^{n} e_k c_k q^{\binom{k}{2}} \, {}^q\nabla e_{n-k} \left[X \frac{1-q^k}{1-q} \right].$$

Garsia and Haiman were also able to obtain the $t = 1/q$ case of ∇e_n, which (using Theorem 2.14) can be expressed as follows.

THEOREM 3.24.

(3.94)
$$q^{\binom{n}{2}} \mathcal{F}(DH_n; q, 1/q) = \frac{1}{[n+1]} e_n \left[X \frac{1-q^{n+1}}{1-q} \right],$$

or, equivalently by the Cauchy identity (1.67),

(3.95)
$$\left\langle q^{\binom{n}{2}} \mathcal{F}(DH_n; q, 1/q), s_\lambda \right\rangle = \frac{1}{[n+1]} s_{\lambda'} \left[\frac{1-q^{n+1}}{1-q} \right].$$

PROOF. The basic reason why ∇e_n can be evaluated when $t = 1/q$ is that after expressing $\tilde{H}[X; q, 1/q]$ in terms of the $P[X; q, t]$, you get a scalar multiple of $P_\mu[\frac{X}{1-q}; q, q]$. But by Remark 2.3, $P_\mu(X; q, q) = s_\mu$.

$\square$

Note that by Theorem 1.15, the special case $\lambda = 1^n$ of (3.95) reduces to (3.39), the formula for MacMahon's maj-statistic q-Catalan.

OPEN PROBLEM 3.25. *Find a q, t-version of the Lagrange inversion formula which will yield an identity for ∇e_n, and which reduces to Theorem 3.22 when $t = 1$ and also incorporates (3.94) when $t = 1/q$.*

CHAPTER 4

The q, t-Schröder Polynomial

The Schröder Bounce and Area Statistics

In this chapter we develop the theory of the q, t-Schröder polynomial, which gives a combinatorial interpretation, in terms of statistics on Schröder lattice paths, for the coefficient of a hook shape in ∇e_n. A *Schröder path* is a lattice path from $(0,0)$ to (n,n) consisting of $N(0,1)$, $E(1,0)$ and diagonal $D(1,1)$ steps which never goes below the line $y = x$. We let $L_{n,n,d}^+$ denote the set of Schröder lattice paths consisting of $n - d$ N steps, $n - d$ E steps, and d D steps. We refer to a triangle whose vertex set is of the form $\{(i,j), (i+1,j), (i+1,j+1)\}$ for some (i,j) as a "lower triangle", and define the area of a Schröder path π to be the number of lower triangles below π and above the line $y = x$. Note that if π has no D steps, then the Schröder definition of area agrees with the definition of the area of a Catalan path. We let $a_i(\pi)$ denote the length of the ith row, from the bottom, of π, so $\text{area}(\pi) = \sum_{i=1}^n a_i(\pi)$.

Given $\pi \in L_{n,n,d}^+$, let $\sigma(\pi)$ be the word of 0's, 1's and 2's obtained in the following way. Initialize σ to be the empty string, then start at $(0,0)$ and travel along π to (n,n), adding a 0, 1, or 2 to the end of $\sigma(\pi)$ when we encounter a N, D, or E step, respectively, of π. (If π is a Dyck path, then this definition of $\sigma(\pi)$ is the same as the previous definition from Chapter 1, except that we end up with a word of 0's and 2's instead of 0's and 1's. Since all our applications involving $\sigma(\pi)$ only depend on the relative order of the elements of $\sigma(\pi)$, this change is only a superficial one.) We define the statistic bounce(π) by means of the following algorithm.

ALGORITHM 4.1.   (1) *First remove all D steps from π, and collapse to obtain a Catalan path $\Gamma(\pi)$. More precisely, let $\Gamma(\pi)$ be the Catalan path for which $\sigma(\Gamma(\pi))$ equals $\sigma(\pi)$ with all 1's removed. Recall the ith peak of $\Gamma(\pi)$ is the lattice point where the bounce path for $\Gamma(\pi)$ switches direction from N to E for the ith time. The lattice point at the beginning of the corresponding E step of π is called the ith peak of π. Say $\Gamma(\pi)$ has b bounce steps.*

(2) *For each D step x of π, let $nump(x)$ be the number of peaks of π, below x. Then define*

(4.1) $$\text{bounce}(\pi) = \text{bounce}(\Gamma(\pi)) + \sum_x nump(x),$$

where the sum is over all D steps of π. For example, if π is the Schröder path on the left in Figure 1, with $\Gamma(\pi)$ on the right, then bounce(π) = $(3 + 1) + (0 + 1 + 1 + 2) = 8$. Note that if π has no D steps, this definition of bounce(π) agrees with the previous definition from Chapter 3.

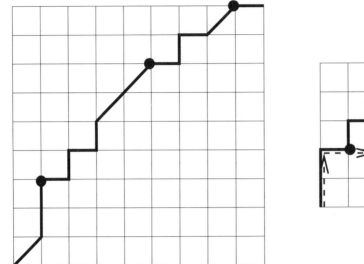

FIGURE 1. On the left, a Schröder path π with the peaks marked by large dots. On the right is $\Gamma(\pi)$ and its bounce path and peaks.

We call the vector whose ith coordinate is the length of the ith bounce step of $\Gamma(\pi)$ the *bounce vector* of π. Say $\Gamma(\pi)$ has b bounce steps. Also, we call the set of rows of π between peaks i and $i+1$ *section i* of π for $1 \leq i \leq b$. In addition we call section 0 the set of rows below peak 1, and section b the set of rows above peak b. If π has β_i D steps in section i, $0 \leq i \leq b$, we refer to $(\beta_0, \beta_1, \ldots, \beta_b)$ as the *shift vector* of π. For example, the path on the left in Figure 1 has bounce vector $(2,2,1)$ and shift vector $(1,2,1,0)$. We refer to the portion of $\sigma(\pi)$ corresponding to the ith section of π as the ith section of $\sigma(\pi)$.

Given $n, d \in \mathbb{N}$, we define the q,t-Schröder polynomial $S_{n,d}(q,t)$ as follows.

$$(4.2) \qquad S_{n,d}(q,t) = \sum_{\pi \in L^+_{n,n,d}} q^{\text{area}(\pi)} t^{\text{bounce}(\pi)}.$$

These polynomials were introduced by Egge, Haglund, Killpatrick and Kremer [**EHKK03**]. They conjectured the following result, which was subsequently proved by Haglund using plethystic results involving Macdonald polynomials [**Hag04b**].

THEOREM 4.2. *For all $0 \leq d \leq n$,*

$$(4.3) \qquad S_{n,d}(q,t) = \langle \nabla e_n, e_{n-d} h_d \rangle.$$

Since

$$(4.4) \qquad S_{n,0}(q,t) = F_n(q,t) = C_n(q,t),$$

the $d = 0$ case of Theorem 4.2 reduces to Theorem 3.2.

Let $\tilde{L}^+_{n,n,d}$ denote the set of paths π which are in $L^+_{n,n,d}$ and also have no D step above the highest N step, i.e. no 1's in $\sigma(\pi)$ after the rightmost 0. Define

$$(4.5) \qquad \tilde{S}_{n,d}(q,t) = \sum_{\pi \in \tilde{L}^+_{n,n,d}} q^{\text{area}(\pi)} t^{\text{bounce}(\pi)}.$$

Then we have

THEOREM 4.3. *Theorem 4.2 is equivalent to the statement that for all $0 \leq d \leq n-1$,*

(4.6)
$$\tilde{S}_{n,d}(q,t) = \left\langle \nabla e_n, s_{d+1,1^{n-d-1}} \right\rangle.$$

PROOF. Given $\pi \in \tilde{L}_{n,n,d}^{+}$, we can map π to a path $\alpha(\pi) \in L_{n,n,d+1}^{+}$ by replacing the highest N step and the following E step of π by a D step. By Exercise 4.4, this map leaves area and bounce unchanged. Conversely, if $\alpha \in L_{n,n,d}^{+}$ has a D step above the highest N step, we can map it to a path $\pi \in \tilde{L}_{n,n,d-1}^{+}$ in an area and bounce preserving fashion by changing the highest D step to a NE pair. It follows that for $1 \leq d \leq n$,

(4.7)
$$S_{n,d}(q,t) = \sum_{\pi \in \tilde{L}_{n,n,d}^{+}} q^{\text{area}(\pi)} t^{\text{bounce}(\pi)} + \sum_{\pi \in \tilde{L}_{n,n,d-1}^{+}} q^{\text{area}(\pi)} t^{\text{bounce}(\pi)}$$
$$= \tilde{S}_{n,d}(q,t) + \tilde{S}_{n,d-1}(q,t).$$

Since $S_{n,0}(q,t) = \tilde{S}_{n,0}(q,t)$, $e_{n-d}h_d = s_{d+1,1^{n-d-1}} + s_{d,1^{n-d}}$ for $0 < d \leq n-1$, and $S_{n,n}(q,t) = 1 = \tilde{S}_{n,n-1}(q,t)$, the result follows by a simple inductive argument. $\square$

EXERCISE 4.4. Given π and $\alpha(\pi)$ in the proof of Theorem 4.3, show that

(4.8)
$$\text{area}(\pi) = \text{area}(\alpha(\pi))$$

(4.9)
$$\text{bounce}(\pi) = \text{bounce}(\alpha(\pi)).$$

Define q,t-analogues of the big Schröder numbers r_n and little Schröder numbers $\tilde{r}_n$ as follows.

(4.10)
$$r_n(q,t) = \sum_{d=0}^{n} S_{n,d}(q,t)$$

(4.11)
$$\tilde{r}_n(q,t) = \sum_{d=0}^{n-1} \tilde{S}_{n,d}(q,t).$$

The numbers $r_n(1,1)$ count the total number of Schröder paths from $(0,0)$ to (n,n). The $\tilde{r}_n(1,1)$ are known to count many different objects [**Sta99**, p.178], including the number of Schröder paths from $(0,0)$ to (n,n) which have no D steps on the line $y = x$. From our comments above we have the simple identity $r_n(q,t) = 2\tilde{r}_n(q,t)$, and using Proposition 2.17.1 and (2.24) we get the polynomial identities

(4.12)
$$\sum_{d=0}^{n} z^d S_{n,d}(q,t) = \sum_{\mu \vdash n} \frac{T_\mu \prod_{x \in \mu}(z + q^{a'} t^{l'}) M \Pi_\mu B_\mu}{w_\mu}$$

(4.13)
$$\sum_{d=0}^{n-1} z^d \tilde{S}_{n,d}(q,t) = \sum_{\mu \vdash n} \frac{T_\mu \prod_{x \in \mu, \, x \neq (0,0)}(z + q^{a'} t^{l'}) M \Pi_\mu B_\mu}{w_\mu}.$$

An interesting special case of (4.13) is

(4.14)
$$\tilde{r}_{n,d}(q,t) = \sum_{\mu \vdash n} \frac{T_\mu \prod'_{x \in \mu}(1 - q^{2a'} t^{2l'}) M B_\mu}{w_\mu}.$$

Recurrences and Explicit Formulae

We begin with a useful lemma about area and Schröder paths.

LEMMA 4.5. *(The "boundary lemma"). Given $a, b, c \in \mathbb{N}$, let boundary(a, b, c) be the path whose σ word is $2^c 1^b 0^a$. Then*

$$(4.15) \qquad \sum_{\pi} q^{area'(\pi)} = \begin{bmatrix} a+b+c \\ a,b,c \end{bmatrix},$$

where the sum is over all paths π from $(0,0)$ to $(c+b, a+b)$ consisting of a N steps, b D steps and c E steps, and area$'(\pi)$ is the number of lower triangles between π and boundary(a, b, c).

PROOF. Given π as above, we claim the number of inversions of rev$(\sigma(\pi))$ equals area$'(\pi)$. To see why, start with π as in Figure 2, and note that when consecutive ND, DE, or NE steps are interchanged, area$'$ decreases by 1. Thus area$'(\pi)$ equals the number of such interchanges needed to transform π into boundary(a, b, c), or equivalently to transform $\sigma(\pi)$ into $2^c 1^b 0^a$. But this is just inv(rev$(\sigma(\pi))$). Thus

$$(4.16) \qquad \sum_{\pi} q^{\text{area}'(\pi)} = \sum_{\sigma \in M_{(a,b,c)}} q^{\text{inv}(\text{rev}(\sigma))}$$

$$= \sum_{\sigma \in M_{(a,b,c)}} q^{\text{inv}(\sigma)}$$

$$= \begin{bmatrix} a+b+c \\ a,b,c \end{bmatrix}$$

by (1.10). $\qquad \square$

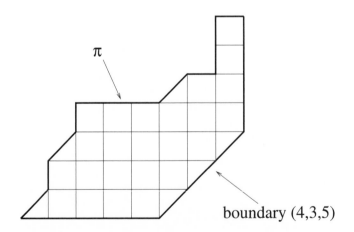

FIGURE 2. The region between a path π and the corresponding boundary path. For this region area$' = 27$.

Given $n, d, k \in \mathbb{N}$ with $1 \le k \le n$, let $L^+_{n,n,d}(k)$ denote the set of paths in $L^+_{n,n,d}$ which have k total D plus N steps below the lowest E step. We view $L^+_{n,n}(k)$ as containing the path with n D steps if $k = n$ and $L^+_{n,n,n}(k)$ as being the emptyset if $k < n$. Define

$$(4.17) \qquad S_{n,d,k}(q,t) = \sum_{\pi \in L^+_{n,n,d}(k)} q^{\text{area}(\pi)} t^{\text{bounce}(\pi)},$$

with $S_{n,n,k}(q,t) = \chi(k = n)$. There is a recursive structure underlying the $S_{n,d,k}(q,t)$ which extends that underlying the $F_{n,k}(q,t)$. The following result is derived in [**Hag04b**], and is similar to recurrence relations occurring in [**EHKK03**]. For any two integers n, k we use the notation

$$(4.18) \qquad \delta_{n,k} = \chi(n = k).$$

THEOREM 4.6. *Let* $n, k, d \in \mathbb{N}$ *with* $1 \le k \le n$. *Then*

$$(4.19) \qquad S_{n,n,k}(q,t) = \delta_{n,k},$$

and for $0 \le d < n$,

$$(4.20) \quad S_{n,d,k}(q,t) = t^{n-k} \sum_{p=\max(1,k-d)}^{\min(k,n-d)} \begin{bmatrix} k \\ p \end{bmatrix} q^{\binom{p}{2}} \sum_{j=0}^{n-k} \begin{bmatrix} p+j-1 \\ j \end{bmatrix} S_{n-k,d+p-k,j}(q,t),$$

with the initial conditions

$$(4.21) \qquad S_{0,0,k} = \delta_{k,0}, \quad S_{n,d,0} = \delta_{n,0}\delta_{d,0}.$$

PROOF. If $d = n$ then (4.19) follows directly from the definition. If $d < n$ then π has at least one peak. Say π has p N steps and $k - p$ D steps in section 0. First assume $p < n - d$, i.e. Π has at least two peaks. We now describe an operation we call truncation, which takes $\pi \in L^+_{n,n,d}(k)$ and maps it to a $\pi' \in L^+_{n-k,n-k,d-k+p}$ with one less peak. Given such a π, to create π' start with $\sigma(\pi)$ and remove the first k letters (section 0). Also remove all the 2's in section 1. The result is $\sigma(\pi')$. For the path on the left in Figure 1, $\sigma(\pi) = 10020201120212$, $k = 3$ and $\sigma(\pi') = 001120212$.

We will use Figure 3 as a visual aid in the remainder of our argument. Let j be the total number of $D + E$ steps of π in section 1 of π. By construction the bounce path for $\Gamma(\pi')$ will be identical to the bounce path for $\Gamma(\Pi)$ except the first bounce of $\Gamma(\pi)$ is truncated. This bounce step hits the diagonal at (p, p), and so the contribution to bounce(π') from the bounce path will be $n - d - p$ less than to bounce(π). Furthermore, for each D step of π above peak 1 of Π, the number of peaks of π' below it will be one less than the number of peaks of π below it. It follows that

$$\text{bounce}(\pi) = \text{bounce}(\pi') + n - d - p + d - (k - p)$$
$$= \text{bounce}(\pi') + n - k.$$

Since the area below the triangle of side p from Figure 3 is $\binom{p}{2}$,

$$(4.22) \qquad \text{area}(\pi) = \text{area}(\pi') + \binom{p}{2} + \text{area0} + \text{area1},$$

where area0 is the area of section 0 of π, and area1 is the portion of the area of section 1 of π not included in area(π'). When we sum $q^{\text{area1}(\pi)}$ over all $\pi \in L^+_{n,n,d}(k)$ which get mapped to π' under truncation, we generate a factor of

$$(4.23) \qquad \begin{bmatrix} k \\ p \end{bmatrix}$$

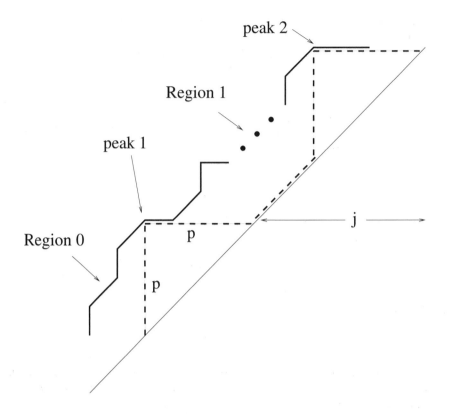

FIGURE 3. A path π decomposed into various regions under truncation.

by the $c = 0$ case of the boundary lemma.

From the proof of the boundary lemma, area1 equals the number of coinversions of the 1st section of $\sigma(\pi)$ involving pairs of 0's and 2's or pairs of 1's and 2's. We need to consider the sum of q to the number of such coinversions, summed over all π which map to π', or equivalently, summed over all ways to interleave the p 2's into the fixed sequence of j 0's and 1's in section 0 of π'. Taking into account the fact that such an interleaving must begin with a 2 but is otherwise unrestricted, we get a factor of

$$(4.24) \qquad \begin{bmatrix} p - 1 + j \\ j \end{bmatrix},$$

since each 2 will form a coinversion with each 0 and each 1 occurring before it. It is now clear how the various terms in (4.20) arise.

Finally, we consider the case when there is only one peak, so $p = n - d$. Since there are $d - (k - p) = d - k + n - d = n - k$ D steps above peak 1 of π, we have bounce$(\pi) = n - k$. Taking area into account, by the above analysis we get

$$(4.25) \qquad S_{n,d,k}(q, t) = t^{n-k} q^{\binom{n-d}{2}} \begin{bmatrix} k \\ n - d \end{bmatrix} \begin{bmatrix} n - d - 1 + n - k \\ n - k \end{bmatrix}$$

which agrees with the $p = n - d$ term on the right-hand-side of (4.20) since $S_{n-k,n-k,j}(q, t) = \delta_{j,n-k}$ from the initial conditions. $\qquad \square$

The following explicit formula for $S_{n,d}(q,t)$ was obtained in [**EHKK03**].

THEOREM 4.7. *For all* $0 \leq d < n$,

$$(4.26) \quad S_{n,d}(q,t) = \sum_{b=1}^{n-d} \sum_{\substack{\alpha_1 + \ldots + \alpha_b = n-d,\, \alpha_i > 0 \\ \beta_0 + \beta_1 + \ldots + \beta_b = d,\, \beta_i \geq 0}} \begin{bmatrix} \beta_0 + \alpha_1 \\ \beta_0 \end{bmatrix} \begin{bmatrix} \beta_b + \alpha_b - 1 \\ \beta_b \end{bmatrix} q^{\binom{\alpha_1}{2} + \ldots + \binom{\alpha_b}{2}}$$

$$t^{\beta_1 + 2\beta_2 + \ldots + b\beta_b + \alpha_2 + 2\alpha_3 + \ldots + (b-1)\alpha_b} \prod_{i=1}^{b-1} \begin{bmatrix} \beta_i + \alpha_{i+1} + \alpha_i - 1 \\ \beta_i, \alpha_{i+1}, \alpha_i - 1 \end{bmatrix}.$$

PROOF. Consider the sum of $q^{\text{area}} t^{\text{bounce}}$ over all π which have bounce vector $(\alpha_1, \ldots, \alpha_b)$, and shift vector $(\beta_0, \beta_1, \ldots, \beta_b)$. For all such π the value of bounce is given by the exponent of t in (4.26). The area below the bounce path generates the $q^{\binom{\alpha_1}{2} + \ldots + \binom{\alpha_b}{2}}$ term. When computing the portion of area above the bounce path, section 0 of π contributes the $\begin{bmatrix} \beta_0 + \alpha_1 \\ \beta_0 \end{bmatrix}$ term. Similarly, section b contributes the $\begin{bmatrix} \beta_b + \alpha_b - 1 \\ \beta_b \end{bmatrix}$ term (the first step above peak b must be an E step by the definition of a peak, which explains why we subtract 1 from α_b). For section i, $1 \leq i < b$, we sum over all ways to interleave the β_i D steps with the α_{i+1} N steps and the α_i E steps, subject to the constraint we start with an E step. By the boundary lemma, we get the $\begin{bmatrix} \beta_i + \alpha_{i+1} + \alpha_i - 1 \\ \beta_i, \alpha_{i+1}, \alpha_i - 1 \end{bmatrix}$ term. $\square$

The Special Value $t = 1/q$

In this section we prove the following result.

THEOREM 4.8. *For* $1 \leq k \leq n$ *and* $0 \leq d \leq n$,

$$(4.27) \quad q^{\binom{n}{2} - \binom{d}{2}} S_{n,d,k}(q, 1/q) = q^{(k-1)(n-d)} \frac{[k]}{[n]} \begin{bmatrix} 2n - k - d - 1 \\ n - k \end{bmatrix} \begin{bmatrix} n \\ d \end{bmatrix}.$$

PROOF. Since this exact result hasn't appeared in print before, we provide a detailed proof, making use of the following identities.

$$(4.28)$$

$$(1 - q^k)(1 - q^{k-1}) \cdots (1 - q^{k-u+1})$$
$$= q^k(1 - q^{-k}) q^{k-1}(1 - q^{1-k}) \cdots q^{k-u+1}(1 - q^{u-k-1})(-1)^u$$
$$= q^{uk - \binom{u}{2}}(-1)^u (q^{-k})_u.$$

$$(4.29) \quad \begin{aligned} \begin{bmatrix} j \\ m - u \end{bmatrix} &= \begin{bmatrix} j \\ m \end{bmatrix} \frac{(1 - q^m)(1 - q^{m-1}) \cdots (1 - q^{m-u+1})}{(q^{j-m+1})_u} \\ &= \begin{bmatrix} j \\ m \end{bmatrix} \frac{(q^{-m})_u (-1)^u q^{mu - \binom{u}{2}}}{(q^{j-m+1})_u}. \end{aligned}$$

$$(4.30) \quad \begin{aligned} \begin{bmatrix} j - u \\ m - u \end{bmatrix} &= \begin{bmatrix} j \\ m \end{bmatrix} \frac{(1 - q^m)(1 - q^{m-1}) \cdots (1 - q^{m-u+1})}{(1 - q^j)(1 - q^{j-1}) \cdots (1 - q^{j-u+1})} \\ &= \begin{bmatrix} j \\ m \end{bmatrix} \frac{(q^{-m})_u q^{mu}}{(q^{-j})_u q^{ju}}. \end{aligned}$$

First assume $k = n$. Then by direct computation,

$$(4.31) \quad S_{n,d,n}(q,t) = q^{\binom{n-d}{2}} \begin{bmatrix} n \\ d \end{bmatrix},$$

so

$$(4.32) \qquad q^{\binom{n}{2}-\binom{d}{2}} S_{n,d,n}(q,1/q) = q^{\binom{n}{2}-\binom{d}{2}+\binom{n-d}{2}} \begin{bmatrix} n \\ d \end{bmatrix}$$

$$= q^{(n-d)(n-1)} \begin{bmatrix} n \\ d \end{bmatrix}$$

which agrees with (4.27).

Next we assume $k < n$, and we apply Theorem 4.20 and induction to get

$$(4.33) \quad q^{\binom{n}{2}-\binom{d}{2}} S_{n,d,k}(q,1/q) = q^{\binom{n}{2}-\binom{d}{2}-(n-k)} \sum_{p=0}^{n-d} \begin{bmatrix} k \\ p \end{bmatrix} q^{\binom{p}{2}}$$

$$\times \sum_{j=0}^{n-k} \begin{bmatrix} p+j-1 \\ j \end{bmatrix} q^{-\binom{n-k}{2}+\binom{d+p-k}{2}} q^{\binom{n-k}{2}-\binom{d+p-k}{2}} S_{n-k,d+p-k,j}(q,1/q)$$

$$= q^{\binom{n}{2}-\binom{d}{2}-(n-k)-\binom{n-k}{2}} \sum_{p=0}^{n-d} \begin{bmatrix} k \\ p \end{bmatrix} q^{\binom{p}{2}} \sum_{j=0}^{n-k} \begin{bmatrix} p+j-1 \\ j \end{bmatrix} q^{(j-1)(n-k-(d+p-k))}$$

$$\times q^{\binom{d+p-k}{2}} \frac{[j]}{[n-k]} \begin{bmatrix} 2(n-k)-j-1-(d+p-k) \\ n-k-j \end{bmatrix} \begin{bmatrix} n-k \\ d+p-k \end{bmatrix}$$

$$(4.34) \quad = q^{\binom{n}{2}-\binom{d}{2}-(n-k)-\binom{n-k}{2}} \frac{1}{[n-k]} \sum_{p=0}^{n-d} \begin{bmatrix} k \\ p \end{bmatrix} q^{\binom{p}{2}+\binom{d+p-k}{2}}$$

$$\times \begin{bmatrix} n-k \\ d+p-k \end{bmatrix} \sum_{\substack{u=0 \\ u=j-1}}^{n-k-1} \begin{bmatrix} p+u \\ u \end{bmatrix} [p] q^{u(n-d-p)} \begin{bmatrix} 2n-k-d-p-2-u \\ n-k-1-u \end{bmatrix}.$$

The inner sum over u in (4.34) equals

$$(4.35)$$

$$[p] \sum_{u=0}^{n-k-1} \begin{bmatrix} p+u \\ u \end{bmatrix} q^{u(n-d-p)} \begin{bmatrix} n-d-p-1+n-k-1-u \\ n-k-1-u \end{bmatrix}$$

$$= [p] \frac{1}{(zq^{n-d-p})_{p+1}(z)_{n-d-p}} |_{z^{n-k-1}}$$

$$= [p] \frac{1}{(z)_{n-d+1}} |_{z^{n-k-1}} = [p] \begin{bmatrix} n-d+n-k-1 \\ n-k-1 \end{bmatrix},$$

using (1.28). Plugging this into (4.34), we now have

$$(4.36) \quad q^{\binom{n}{2}-\binom{d}{2}} S_{n,d,k}(q,1/q) = q^{\binom{n}{2}-\binom{d}{2}-(n-k)-\binom{n-k}{2}} \frac{1}{[n-k]} \begin{bmatrix} 2n-k-d-1 \\ n-k-1 \end{bmatrix}$$

$$\times \sum_{p=0}^{n-d} \begin{bmatrix} k \\ p \end{bmatrix} [p] q^{\binom{p}{2}+\binom{d+p-k}{2}} \begin{bmatrix} n-k \\ d+p-k \end{bmatrix}.$$

The sum over p equals

$$(4.37) \quad \sum_{\substack{u=0 \\ u=p-1}}^{n-d-1} \begin{bmatrix} k-1 \\ u \end{bmatrix} [k] q^{\binom{u+1}{2} + \binom{d-k+1+u}{2}} \begin{bmatrix} n-k \\ d-k+1+u \end{bmatrix}$$

$$= [k] \sum_{u=0}^{n-d-1} \begin{bmatrix} k-1 \\ u \end{bmatrix} q^{\binom{u}{2} + u + \binom{d-k+1+u}{2}} \begin{bmatrix} n-k \\ n-d-1-u \end{bmatrix}$$

$$= [k] \sum_{u=0}^{n-d-1} \begin{bmatrix} k-1 \\ u \end{bmatrix} q^{\binom{u}{2} + u} \begin{bmatrix} n-k \\ n-d-1-u \end{bmatrix} q^{\binom{n-d-1-u}{2}} q^{\text{pow}},$$

where

$$\text{pow} = \binom{d-k+1+u}{2} - \binom{n-d-1-u}{2}$$

$$= \binom{d-k+1}{2} + \binom{u}{2} + (d-k+1)u - \binom{n-d-1}{2} + \binom{u}{2} + u(n-d-1-u)$$

$$= \binom{d-k+1}{2} + 2\binom{u}{2} + (d-k+1+n-d-1)u - u^2 - \binom{n-d-1}{2}$$

$$= \binom{d-k+1}{2} - \binom{n-d-1}{2} + (n-k-1)u.$$

Thus the sum over p in (4.36) equals

$$(4.38) \quad [k] q^{\binom{d-k+1}{2} - \binom{n-d-1}{2}} \sum_{u=0}^{n-d-1} \begin{bmatrix} k-1 \\ u \end{bmatrix} q^{\binom{u}{2} + u(n-k)} \begin{bmatrix} n-k \\ n-d-1-u \end{bmatrix} q^{\binom{n-d-1-u}{2}}$$

$$= [k] q^{\binom{d-k+1}{2} - \binom{n-d-1}{2}} (-zq^{n-k})_{k-1} (-z)_{n-k} \big|_{z^{n-1-d}}$$

$$= [k] q^{\binom{d-k+1}{2} - \binom{n-d-1}{2}} (-z)_{n-1} \big|_{z^{n-1-d}}$$

$$= [k] q^{\binom{d-k+1}{2} - \binom{n-d-1}{2}} \begin{bmatrix} n-1 \\ d \end{bmatrix} q^{\binom{n-d-1}{2}}.$$

Plugging this into (4.36) we finally obtain

$$S_{n,d,k}(q, 1/q) = q^{-(n-k) - \binom{n-k}{2}} \frac{1}{[n-k]} \begin{bmatrix} 2n-k-d-1 \\ n-k-1 \end{bmatrix} [k] q^{\binom{d-k+1}{2}} \begin{bmatrix} n-1 \\ d \end{bmatrix}$$

$$= \frac{[k]}{[n]} \begin{bmatrix} 2n-k-1-d \\ n-k \end{bmatrix} \begin{bmatrix} n \\ d \end{bmatrix} q^{\text{pow}},$$

where

$$\text{pow} = \binom{n}{2} - \binom{d}{2} - n + k - \binom{n-k}{2} + \binom{d-k+1}{2}$$

$$= \frac{n^2 - n}{2} - \frac{d^2 - d}{2} - n + k - \frac{(n-k)(n-k-1)}{2} + \frac{(d-k+1)(d-k)}{2}$$

$$= d - dk - n + nk = (k-1)(n-d).$$

$\square$

COROLLARY 4.8.1. *For* $0 \leq d \leq n$,

$$(4.39) \quad q^{\binom{n}{2} - \binom{d}{2}} S_{n,d}(q, 1/q) = \frac{1}{[n-d+1]} \begin{bmatrix} 2n-d \\ n-d, n-d, d \end{bmatrix}.$$

PROOF. It is easy to see combinatorially that $S_{n+1,d,1}(q,t) = t^n S_{n,d}(q,t)$. Thus

$$(4.40) \quad q^{\binom{n}{2}-\binom{d}{2}} S_{n,d}(q,1/q) = q^{\binom{n}{2}+n-\binom{d}{2}} S_{n+1,d,1}(q,1/q)$$

$$(4.41) \qquad\qquad = q^{\binom{n+1}{2}-\binom{d}{2}} S_{n+1,d,1}(q,1/q)$$

$$(4.42) \qquad\qquad = q^{(1-1)(n+1-d)} \frac{[1]}{[n+1]} \left[\begin{matrix} 2(n+1)-1-d-1 \\ n \end{matrix} \right] \left[\begin{matrix} n+1 \\ d \end{matrix} \right]$$

$$(4.43) \qquad\qquad = \frac{1}{[n-d+1]} \left[\begin{matrix} 2n-d \\ n-d, n-d, d \end{matrix} \right].$$

$\square$

REMARK 4.9. Corollary 4.8.1 proves that $S_{n,d}(q,t)$ is symmetric in q,t when $t = 1/q$. For we have

$$(4.44) \qquad S_{n,d}(q,1/q) = q^{-\binom{n}{2}+\binom{d}{2}} \frac{1}{[n-d+1]} \left[\begin{matrix} 2n-d \\ n-d, n-d, d \end{matrix} \right],$$

and replacing q by $1/q$ we get

$$(4.45) \quad S_{n,d}(1/q,q) = q^{\binom{n}{2}-\binom{d}{2}} \frac{q^{n-d}}{[n-d+1]} \left[\begin{matrix} 2n-d \\ n-d, n-d, d \end{matrix} \right] q^{2\binom{n-d}{2}+\binom{d}{2}-\binom{2n-d}{2}},$$

since $[n]!_{1/q} = [n]! q^{-\binom{n}{2}}$. Now

$$(4.46) \quad \binom{n}{2} - \binom{d}{2} + n - d + 2 \binom{n-d}{2} + \binom{d}{2} - \binom{2n-d}{2} = \binom{d}{2} - \binom{n}{2},$$

so $S_{n,d}(q,1/q) = S_{n,d}(1/q,q)$. It is of course an open problem to show $S_{n,d}(q,t) = S_{n,d}(t,q)$ bijectively, since the $d = 0$ case is Open Problem 3.11.

The Delta Operator

We begin this section with a refinement of Theorem 4.2 [**Hag04b**]. If $d = 0$ it reduces to (3.26), the formula for $F_{n,k}(q,t)$ in terms of the $E_{n,k}$.

THEOREM 4.10. *For all* $1 \le k \le n$ *and* $0 \le d \le n$,

$$(4.47) \qquad\qquad \langle \nabla E_{n,k}, e_{n-d} h_d \rangle = S_{n,d,k}(q,t).$$

PROOF. The proof, which is presented in Chapter 7, uses the same method as the proof of Corollary 3.5.1, by showing the left-hand side of (4.47) satisfies the same recurrence and initial conditions as $S_{n,d,k}(q,t)$. $\square$

There is another formula for $S_{n,d,k}(q,t)$, which involves a linear operator Δ depending on an arbitrary symmetric function f, which is defined on the $\tilde{H}_\mu$ basis via

$$(4.48) \qquad\qquad \Delta_f \tilde{H}_\mu = f[B_\mu] \tilde{H}_\mu.$$

The Δ operator occurs often in the theory of diagonal harmonics. In particular we have the following important result of Haiman [**Hai02**], who shows the polynomials below have a geometric interpretation.

THEOREM 4.11. *For all* $\lambda, \beta \in Par$,

$$(4.49) \qquad\qquad \langle \Delta_{s_\lambda} \nabla e_n, s_\beta \rangle \in \mathbb{N}[q,t].$$

In [**Hag04b**] the following conjectured refinement of the nonnegativity of the polynomials above is proposed.

CONJECTURE 4.12. *For all* $1 \leq k \leq n$,

$$(4.50) \qquad \langle \Delta_{s_\lambda} \nabla E_{n,k}, s_\beta \rangle \in \mathbb{N}[q,t].$$

REMARK 4.13. It has been conjectured [**BGHT99**, p.419] that

$$(4.51) \qquad \langle \Delta_{s_\lambda} e_n, s_\beta \rangle \in \mathbb{N}[q,t]$$

(note the lack of ∇ before the e_n). This stronger conjecture is false with e_n replaced by $E_{n,k}$. In particular, in general $\langle E_{n,k}, s_\beta \rangle \notin \mathbb{N}[q,t]$.

From [**Hag04b**] we have

THEOREM 4.14. *For* $1 \leq k \leq n$,

$$(4.52) \qquad \langle \nabla E_{n,k}, s_n \rangle = \chi(n = k),$$

and in addition, if $|\lambda| = m > 0$,

$$(4.53) \qquad \langle \Delta_{s_\lambda} \nabla E_{n,k}, s_n \rangle = t^{n-k} \left\langle \Delta_{h_{n-k}} e_m \left[X \frac{1-q^k}{1-q} \right], s_{\lambda'} \right\rangle,$$

or equivalently, by (3.18),

$$(4.54) \qquad \langle \Delta_{s_\lambda} \nabla E_{n,k}, s_n \rangle = t^{n-k} \sum_{\mu \vdash m} \frac{(1-q^k) h_k[(1-t)B_\mu] h_{n-k}[B_\mu] \Pi_\mu \tilde{K}_{\lambda',\mu}}{w_\mu}.$$

COROLLARY 4.14.1. *For* $1 \leq k \leq n$ *and* $0 \leq d \leq n$,

$$(4.55) \qquad S_{n,d,k}(q,t) = t^{n-k} \left\langle \Delta_{h_{n-k}} e_{n-d} \left[X \frac{1-q^k}{1-q} \right], s_{n-d} \right\rangle.$$

COROLLARY 4.14.2. *For* $0 \leq d \leq n$,

$$(4.56) \qquad S_{n,d}(q,t) = \langle \Delta_{h_n} e_{n+1-d}, s_{n+1-d} \rangle.$$

When $d = 0$, Corollaries 4.14.1 and 4.14.2 reduce to new formulas for the q,t-Catalan and its stratified version $F_{n,k}(q,t)$.

EXERCISE 4.15. Show that Corollary 4.14.1 follows from (4.47), Theorem 4.14 and (2.24), and also that Corollary 4.14.2 is a special case of Corollary 4.14.1.

The following general identity is a special case of a result from [**Hag04b**] that will be discussed in more detail in Chapter 7.

THEOREM 4.16. *Given positive integers* n, m, $\lambda \vdash m$, $z \in \mathbb{R}$, *and a symmetric function* $P \in \Lambda^n$,

$$(4.57) \qquad \sum_{p=1}^{m} \begin{bmatrix} z \\ p \end{bmatrix} q^{\binom{p}{2}} \left\langle \Delta_{e_{m-p}} e_n \left[X \frac{1-q^p}{1-q} \right], P \right\rangle = \left\langle \Delta_{\omega P} e_m \left[X \frac{1-q^z}{1-q} \right], s_m \right\rangle.$$

We now list some interesting consequences of Theorem 4.16. If $z = 1$ we get

COROLLARY 4.16.1. *Given positive integers* n, m, *and* $P \in \Lambda^n$,

$$(4.58) \qquad \langle \Delta_{e_{m-1}} e_n, P \rangle = \langle \Delta_{\omega P} e_m, s_m \rangle.$$

EXAMPLE 4.17. Letting $m = n + 1$ and $P = h_{1^n}$ in Corollary 4.16.1 we get the following formula for the Hilbert series for DH_n.

$$(4.59) \qquad \mathcal{H}(DH_n; q, t) = \sum_{\mu \vdash n+1} \frac{M(B_\mu)^{n+1} \Pi_\mu}{w_\mu}.$$

Similarly, we have

$$(4.60) \qquad \langle \nabla e_n, s_\lambda \rangle = \langle \Delta_{s_{\lambda'}} e_{n+1}, s_{n+1} \rangle$$
$$= \sum_{\mu \vdash n+1} \frac{M B_\mu s_{\lambda'}[B_\mu] \Pi_\mu}{w_\mu},$$

which gives an expression for the Schur coefficients of ∇e_n without any reference to Macdonald polynomials.

The Schröder dinv Statistic

Given $\pi \in L^+_{n,n,d}$ and $1 \leq i \leq n$, we call row_i of π an N row if row_i contains an N step of π, otherwise we call row_i a D row.

DEFINITION 4.18. Let $\text{dinv}(\pi)$ be the number of pairs (i, j), $1 \leq i < j \leq n$, such that either

(1) $a_i(\pi) = a_j(\pi)$ and row_i is an N row
   or
(2) $a_i(\pi) = a_j(\pi) + 1$ and row_j is an N row.

For example, for the path on the left side of Figure 4, the inversion pairs are $(3, 4)$, $(1, 6)$, $(2, 6)$, $(3, 6)$, $(4, 6)$, $(3, 7)$, $(4, 7)$, $(5, 7)$, $(3, 8)$, $(4, 8)$, $(5, 8)$, $(7, 8)$ and $\text{dinv}(\pi) = 12$.

We call the length of the longest N-row of π the *height* of π. For π of height h, we call the vector $(n_0, n_1, \ldots, n_h)$ whose ith coordinate is the number of N rows of Π of length $i - 1$ the N-area vector of Π. Similarly we call the vector $(d_0, d_1, \ldots, d_{h+1})$ whose ith coordinate is the number of D rows of π of length $i - 1$ the D-area vector of Π.

The following result is taken from [**EHKK03**].

THEOREM 4.19. *Given $\pi \in L^+_{n,n,d}$, the following "sweep" algorithm is a bijective map which creates a path $\phi(\pi) \in L^+_{n,n,d}$ with the following properties.*

$$(4.61) \qquad dinv(\pi) = area(\phi(\pi)) \text{ and } area(\pi) = bounce(\phi(\pi)).$$

Moreover, the N-area vector of π equals the bounce vector of $\phi(\pi)$, and the D-area vector of π equals the shift vector of $\phi(\pi)$.

Algorithm ϕ [(dinv, area) $\rightarrow$ (area, bounce)]:
Input: the path π of height h.
We create the path $\phi(\pi)$ by starting at the lattice point (n, n) at the end of ϕ and appending steps one by one, ending at $(0, 0)$.
Initialize ϕ to $\emptyset$.
For $v = h$ to -1;
    For $w = n$ to 1;
        If $a_w(\pi) = v$ and row_w is an N row then append an E step to ϕ;
        If $a_w(\pi) = v + 1$ and row_w is a D row then append a D step to ϕ;
        If $a_w(\pi) = v + 1$ and row_w is an N row then append an N step to ϕ;

repeat;

repeat;

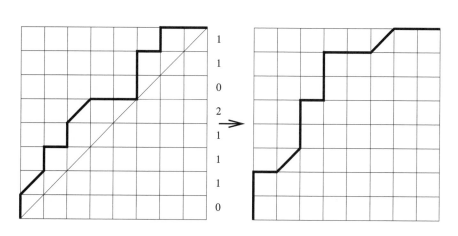

FIGURE 4. On the left, a path π and on the right, $\phi(\pi)$.

PROOF. As we proceed through the proof, the reader may wish to consult Figure 4, which gives an example of the ϕ map applied to the path on the left. First note that if row_i is an N row with length j, then it will first contribute an E step to ϕ during the $v = j$ loop, then an N step to ϕ during the $v = j - 1$ loop. If row_i is a D row of length j, then it will contribute a D step to ϕ during the $v = j - 1$ loop. It follows that $\phi(\pi) \in L_{n,n,d}^+$.

We now prove by induction on i that for $0 \leq i \leq h$, the last step created during the $v = i$ loop is an E step just to the right of peak $i + 1$ of ϕ, and also that the $i + 1$st coordinates of the bounce and shift vectors of ϕ equal n_i and d_i, respectively, the $i + 1$st coordinates of the N-area and D-area vectors of π. The $v = -1$ loop creates n_0 N steps and d_0 D steps. Now it is clear geometrically that for any π, if $\text{area}_j = v + 1$ for $v > -1$ then there exists $l < j$ such that $\text{area}_l = i$ and row_l is an N row. Thus for $i > -1$ the last step of $\phi(\pi)$ created during the $v = i$ loop will be an E step. In particular, the last step created during the $v = 0$ loop is the E step immediately to the right of peak 1 of $\phi(\pi)$. Now assume by induction that the last step created during the $v = i - 1$ loop is an E step immediately to the right of peak i, and that the length of the ith bounce step of $\phi(\pi)$ is n_{i-1}. The fact that the last step created during the $v = i$ loop is an E step, together with the fact that the number of E steps created during the $v = i - 1$ loop is n_{i-1}, imply that the length of the $i + 1$st bounce step of $\phi(\pi)$ equals the number of N steps created during the $v = i - 1$ loop, which is n_i. Also, the last E step created during the $v = i$ loop will be immediately to the right of peak $i + 1$. Furthermore, the number of D steps created during the $v = i - 1$ loop is d_i, and these are exactly the set of D steps in section i of $\phi(\pi)$. This completes the induction.

Now $\text{area}(\phi(\pi)) = \sum_i \binom{n_i}{2}$, plus the area in each of the sections of $\phi(\pi)$ between the path and the corresponding boundary path. Note $\sum_i \binom{n_i}{2}$ counts the number of inversion pairs of π between rows of the same length, neither of which is a D row. Fix i, $0 \leq i \leq k$, and let $\text{row}_{j_1}, \text{row}_{j_2}, \ldots, \text{row}_{j_p}$, $j_1 > j_2 > \cdots > j_p$ be the sequence of rows of π affecting the $v = i$ loop, i.e. rows which are either N rows of length i

or $i+1$, or D rows of length $i+1$ (so $p = n_i + n_{i+1} + d_{i+1}$). Let τ be the subword of $\sigma(\pi)$ consisting of of 2's, 1's and 0's corresponding to the portion of π affecting the $v = i$ loop. The definition of dinv implies the number of inversion pairs of π of the form (x, y) with row$_x$ of length i, row$_y$ of length $i+1$ and x an N row equals the number of coinversion pairs of τ involving 2's and 1's or 2's and 0's. Similarly, the number of inversion pairs of π of the form (x, y) with row$_x$ of length $i+1$, row$_y$ of length $i+1$ and y an N row equals the number of coinversion pairs of τ involving 1's and 0's. By the proof of the boundary lemma, dinv$(\pi) = $ area$(\phi(\pi))$.

It remains to show the algorithm is invertible. The bounce and shift vectors of $\phi(\pi)$ yield the N-area and D-area vectors of π, respectively. In particular they tell us how many rows of π there are of length 0. From section 0 of $\phi(\pi)$ we can determine which subset of these are D rows, since the $v = -1$ iteration of the ϕ algorithm produces an E step in section 0 of $\phi(\pi)$ for each N row of length 0 in π, and a D step in section 0 of $\phi(\pi)$ for each D row of length 0 in π. From section 1 of $\phi(\pi)$ we can determine how the rows of length 1 of π are interleaved with the rows of length 0 of π, and also which rows of length 1 are D rows, since the $v = 0$ iteration of ϕ creates an N step in section 1 of $\phi(\pi)$ for every N row of π of length 0, and a D or E step of $\phi(\pi)$ for every N-row or D row, respectively, of length 1 in π. When considering how the rows of length 2 of π are related to the rows of length 1, we can ignore the rows of length 0, since no row of length 0 can be directly below a row of length 2 and still be the row sequence of a Schröder path. Hence from section 2 of $\phi(\pi)$ we can determine how the rows of length 2 of π are interleaved with the rows of length 1, and which ones are D rows. Continuing in this way we can completely determine π. Below is an explicit algorithm for the inverse.

Algorithm ϕ^{-1} [(area, bounce) $\rightarrow$ (dinv, area)]:
Initialize to $(a) = (-1, N)$.
Input: a path $\phi(\pi) \in S_{n,d}$ with b peaks, where the top of peak i has coordinates (x_i, y_i) for $1 \le i \le b$. Define $x_{b+1} = n$, $y_{b+1} = n$ and $(x_0, y_0) = (0, 0)$.
Let M_i denote the number of steps in the ith section of $\phi(\pi)$, $0 \le i \le b$.
Output: a sequence of pairs $(a_i(\pi), R_i)_{i=1}^n$, where R_i is either N or D indicating whether row$_i$ of π is an N row or a D row.
For $i = 1$ to $b + 1$;
 Number the steps of $\phi(\pi)$ beginning at (x_i, y_i), moving down the path
 until reaching (x_{i-1}, y_{i-1}).
 Given the sequence (a) created thus far, we insert a new subsequence of
 $(i - 1, N)$'s and $(i - 1, D)$'s starting to the left of the first element of (a)
 and moving to the right.
 For $j = 1$ to M_{i-1};
  If step j is an E step then move to the right past the next $(i - 2, N)$ in (a);
  If step j is a D step then insert a $(i - 1, D)$
  immediately to the left of the next $(i - 2, N)$ in (a);
  If step j is an N step then insert a $(i - 1, N)$
  immediately to the left of the next $(i - 2, N)$ in (a);
 repeat;
repeat;
remove the $(-1, N)$ from (a), then reverse (a).

$\square$

COROLLARY 4.19.1. *For $0 \le d \le n$,*

$$(4.62) \qquad S_{n,d,k}(q,t) = \sum_{\substack{\pi \in L^+_{n,n,d} \\ k \text{ rows of length } 0}} q^{dinv(\pi)} t^{area(\pi)},$$

where the sum is over all $\pi \in L^+_{n,n,d}$ having exactly k rows of length 0.

COROLLARY 4.19.2. *For $0 \le d \le n$,*

$$(4.63) \qquad S_{n,d}(q,1) = S_{n,d}(1,q).$$

EXERCISE 4.20. Generalize Exercise 3.18 by explaining why the right-hand-side of (4.26) equals the sum of $q^{\mathrm{dinv}(\pi)} t^{\mathrm{area}(\pi)}$ over $\pi \in L^+_{n,n,d}$, without referencing any results involving the bounce statistic.

The Limit as $d \to \infty$

In this section we discuss the following limiting case of the q,t-Schröder polynomial, which will allow us to express the right-hand-side below as an infinite sum over Schröder paths.

THEOREM 4.21. *For $n \in \mathbb{N}$,*

$$(4.64) \qquad \lim_{d \to \infty} S_{n+d,d}(q,t) = \prod_{i,j \ge 0} (1 + q^i t^j z)|_{z^n}.$$

PROOF. By Corollary 4.14.2 we have

$$(4.65) \qquad S_{n+d,d}(q,t) = \sum_{\mu \vdash n+1} \frac{MB_\mu h_{n+d}[B_\mu] \Pi_\mu}{w_\mu}.$$

Now

$$(4.66) \qquad \lim_{d \to \infty} h_{n+d}[B_\mu] = \lim_{d \to \infty} \prod_{(i,j) \in \mu} \frac{1}{1 - q^{a'} t^{l'} z}|_{z^{n+d}}$$

$$(4.67) \qquad = \lim_{d \to \infty} \frac{1}{1 - z} \prod_{\substack{(i,j) \in \mu \\ (i,j) \ne (0,0)}} \frac{1}{1 - q^{a'} t^{l'} z}|_{z^{n+d}}$$

$$(4.68) \qquad = \prod_{(i,j) \in \mu, (i,j) \ne (0,0)} \frac{1}{1 - q^{a'} t^{l'}} = \frac{1}{\Pi_\mu}.$$

Thus

$$(4.69) \qquad \lim_{d \to \infty} S_{n+d,d}(q,t) = \sum_{\mu \vdash n+1} \frac{MB_\mu}{w_\mu}$$

$$(4.70) \qquad = M \left\langle e_{n+1} \left[\frac{X}{M} \right], e_1 h_n \right\rangle$$

by the $Y = 1$ case of (2.68), the fact that $\tilde{H}_\mu[1; q, t] = 1$ for all μ, and (2.24). Using (1.68) in (4.70) we get

$$(4.71) \qquad \lim_{d \to \infty} S_{n+d,d}(q,t) = M \sum_{\lambda \vdash n+1} \left\langle s_\lambda[X] s_{\lambda'} \left[\frac{1}{M}\right], e_1 h_n \right\rangle$$

$$(4.72) \qquad\qquad = M h_1 \left[\frac{1}{M}\right] e_n \left[\frac{1}{M}\right]$$

$$(4.73) \qquad\qquad = e_n \left[\frac{1}{M}\right].$$

Now by (1.35)

$$(4.74) \qquad \prod_{i,j \geq 0} (1 + q^i t^j z)|_{z^n} = e_n \left[\frac{1}{M}\right],$$

and the result follows. $\qquad\square$

COROLLARY 4.21.1. *For $|q|, |t| < 1$,*

$$(4.75)$$
$$\prod_{i,j \geq 0} (1 + q^i t^j z)|_{z^n} = \sum_{d=0}^{\infty} \sum_{\pi \in L^+_{n+d,n+d,d},\, \sigma_1=0} q^{area(\pi)} t^{bounce(\pi)}$$

$$(4.76) \qquad = \sum_{\substack{\pi \text{ has } n \text{ N, E steps, any } \# \text{ of D steps}, \sigma_1=0}} q^{area(\pi)} t^{bounce(\pi)},$$

where in the sums above we require that π begins with an N step, i.e. $\sigma_1 = 0$.

PROOF. Note that if we add a D step to the beginning of a path π, area(π) and bounce(π) remain the same. We fix the number of N and E steps to be n, and require d D steps, and break up our paths according to how many D steps they begin with. After truncating these beginning D steps, we get

$$(4.77) \qquad S_{n+d,d}(q,t) = \sum_{p=0}^{d} \sum_{\substack{\pi \in L^+_{n+p,n+p,p} \\ \pi \text{ begins with an } N \text{ step}}} q^{area(\pi)} t^{bounce(\pi)}.$$

Assuming $|q|, |t| < 1$ (4.74) implies (4.77) converges absolutely as $d \to \infty$, so it can be rearranged to the form (4.76). $\qquad\square$

OPEN PROBLEM 4.22. *Prove Corollary 4.21.1 bijectively.*

EXERCISE 4.23. Derive the following formula, for the limit as $d \to \infty$ of $S_{n+d,d}(q,t)$, by starting with Theorem 4.7, replacing n by $n + d$ and then taking the limit as $d \to \infty$.

$$(4.78) \qquad \sum_{b=1}^{n} \sum_{\alpha_1 + \ldots + \alpha_b = n,\, \alpha_i > 0} q^{\sum_{i=1}^{b} \binom{\alpha_i}{2}} t^{\sum_{i=2}^{b} (i-1)\alpha_i} \frac{1}{(t^b; q)_{\alpha_b} (q)_{\alpha_1}}$$

$$\times \prod_{i=1}^{b-1} \begin{bmatrix} \alpha_i + \alpha_{i+1} - 1 \\ \alpha_{i+1} \end{bmatrix} \frac{1}{(t^i; q)_{\alpha_i + \alpha_{i+1}}}.$$

Various special cases of (4.75) are proved bijectively by Chunwei Song in his Ph.D. thesis [**Son04**]. He also advances some associated conjectures, as well as studying Schröder paths for a $nm \times n$ rectangle. It is hoped that a bijective proof that (4.75) or (4.78) equals $\prod_{i,j \geq 0}(1 + q^i t^j)|_{z^n}$ would in turn give some insight into Problem 3.11.

CHAPTER 5

Parking Functions and the Hilbert Series

Extension of the dinv Statistic

Recall that the dimension of DH_n equals $(n+1)^{n-1}$, which is the number of *parking functions* on n cars. These functions can be represented geometrically, by starting with a Dyck path π and then forming a parking function P by placing the numbers, or "cars" 1 through n in the squares just to the right of N steps of π, with strict decrease down columns. To see why P is called a parking function, imagine a one way street below the path π with n parking spots $1, 2, \ldots, n$ below columns $\{1, 2, \ldots, n\}$, as in Figure 1. By column i we mean the ith column from the left. Each car has a preferred spot to park in, with the cars in column 1 all wanting to park in spot 1, and more generally the cars in column j, if any, all want to park in spot j, $1 \le j \le n$. Car 1 gets to park first, and parks in its preferred spot. Car 2 gets to park next, and parks in its preferred spot if it is still available. If not, it drives to the right and parks in the first available spot. Then car 3 tries to park in its preferred spot, traveling to the right to find the first available spot if necessary. We continue in this way until all cars park. We call the permutation obtained by reading the cars in spots $1, 2, \ldots, n$ in order the *parking order* of P.

It is easy to see that the fact that π never goes below the line $y = x$ guarantees that all the cars get to park, hence the term "parking function". We can also view P as a "preference function" $f : \{1, \ldots, n\} \to \{1, \ldots, n\}$, where $f(i) = j$ means car i prefers to park in spot j. For such an f we must have the $|f^{-1}(\{1, \ldots, i\})| \ge i$ for $1 \le i \le n$, i.e. for each i at least i cars prefer to park in spots 1 through i, which is guaranteed since π is a Dyck path.

We have the following well-known result.

PROPOSITION 5.0.1. *The number of parking functions on n cars is $(n+1)^{n-1}$.*

PROOF. Imagine that the one way street is circular, with $n+1$ spots, and that each car can prefer to park in any of the $n+1$ spots. There are thus $(n+1)^n$ possible such "generic choice functions". As before, car 1 begins and parks in its preferred spot. Car 2 then parks, driving clockwise around the circular street if necessary to find the first available spot. Because the street is circular, all cars will get to park and moreover there will be exactly one open spot at the end. The generic choice function equals a parking function if and only if the empty spot is spot $n+1$. By symmetry, the number of generic choice functions where spot i ends up empty is the same for all i, thus the number of parking functions is $(n+1)^n/(n+1)$. □

The number $(n+1)^{n-1}$ is also known to count the number of rooted, labeled trees on $n+1$ vertices with root node labeled 0. The set of such trees for $n = 2$ is shown in Figure 2. We assume that all the labels on the children of any given node increase from left to right.

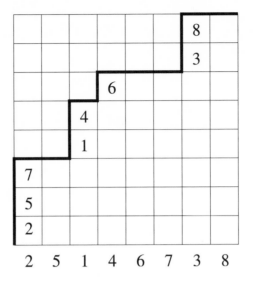

FIGURE 1. A parking function P with parking order 25146738.

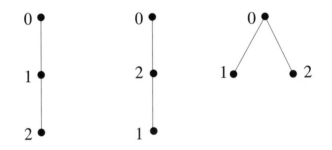

FIGURE 2. The rooted, labeled trees for $n = 2$.

There are a number of bijections between parking functions and trees. One which occurs in [**HL05**] identifies car i with a node with label i in the following simple way. Let the cars in column 1 be the children of the root node. For any other node with label i, to determine its children start at car i and travel NE at a 45 degree angle, staying in the same diagonal, until you either bump into another car or else leave the grid. If you bump into another car, and that car is at the bottom of a column, then all the cars in that column are the children of node i, else node i has no children. For example, the parking function in Figure 1 translates into the tree in Figure 3.

We now describe an extension of the dinv statistic to parking functions. Let $\mathcal{P}_n$ denote the set of all parking functions on n cars. Given $P \in \mathcal{P}_n$ with associated Dyck path $\pi = \pi(P)$, if car i is in row j we say occupant$(j) = i$. Let dinv(P) be the number of pairs (i, j), $1 \le i < j \le n$ such that

$$\text{dinv}(P) = |\{(i,j) : 1 \le i < j \le n \quad a_i = a_j \text{ and occupant}(i) < \text{occupant}(j)\}|$$
$$+ |\{(i,j) : 1 \le i < j \le n \quad a_i = a_j + 1 \text{ and occupant}(i) > \text{occupant}(j)\}|.$$

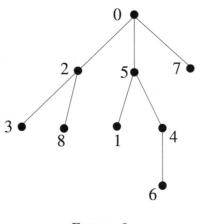

FIGURE 3

Thus $\operatorname{dinv}(P)$ is the number of pairs of rows of P of the same length, with the row above containing the larger car, or which differ by one in length, with the longer row below the shorter, and the longer row containing the larger car. For example, for the parking function in Figure 1, the inversion pairs (i,j) are $(1,7)$, $(2,7)$, $(2,8)$, $(3,4)$, $(4,8)$ and $(5,6)$, so $\operatorname{dinv}(P) = 6$.

We define $\operatorname{area}(P) = \operatorname{area}(\pi)$, and also define the *reading word* of P, denoted $\operatorname{read}(P)$, to be the permutation obtained by reading the cars along diagonals in a SE direction, starting with the diagonal farthest from the line $y = x$, then working inwards. For example, the parking function in Figure 1 has reading word 64781532.

REMARK 5.1. Note that $\operatorname{read}(P) = n \cdots 21$ if and only if $\operatorname{dinv}(P) = \operatorname{dinv}(\pi)$. We call this parking function the Maxdinv parking function for π, which we denote by $\operatorname{Maxdinv}(\pi)$.

Recall that Proposition 2.17.1 implies

(5.1) $$\mathcal{H}(DH_n; q, t) = \langle \nabla e_n, h_{1^n} \rangle.$$

In [**HL05**] N. Loehr and the author advance the following conjectured combinatorial formula for the Hilbert series, which is still open.

CONJECTURE 5.2.

(5.2) $$\langle \nabla e_n, h_1^n \rangle = \sum_{P \in \mathcal{P}_n} q^{\operatorname{dinv}(P)} t^{\operatorname{area}(P)}.$$

Conjecture 5.2 has been verified in Maple for $n \leq 11$. The truth of the conjecture when $q = 1$ follows from results of Garsia and Haiman in [**GH96a**]. Later in the chapter we will show (Corollary 5.6.1) that dinv has the same distribution as area over $\mathcal{P}_n$, which implies the conjecture is also true when $t = 1$.

An Explicit Formula

Given $\tau \in S_n$, with descents at places $i_1 < i_2 < \ldots < i_k$, we call the first i_1 letters of τ the first run of τ, the next $i_2 - i_1$ letters of τ the second run of τ, $\ldots$, and the last $n - i_k$ letters of τ the $(k+1)$st run of τ. For example, the runs of 58246137 are 58, 246 and 137. It will prove convenient to call element 0 the $k+2$nd run of τ. Let $\operatorname{cars}(\tau)$ denote the set of parking functions whose cars in rows of length 0 consist of the elements of the $(k+1)$st run of τ (in any order), whose cars in rows of length 1 consist of the elements of the kth run of τ (in any order), ..., and whose cars in rows of length k consist of the elements of the first run of τ (in any order). For example, the elements of cars(31254) are listed in Figure 4.

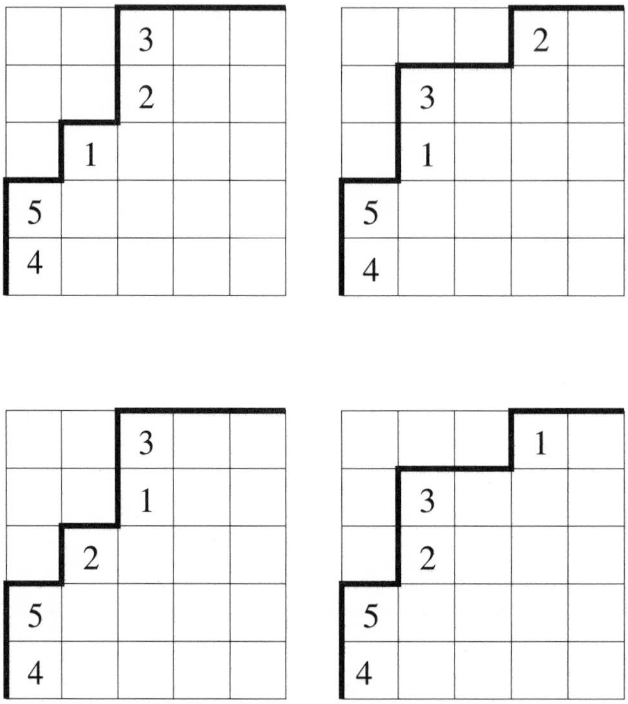

FIGURE 4

Let τ be as above, and let i be a given integer satisfying $1 \le i \le n$. If τ_i is in the jth run of τ, we define $w_i(\tau)$ to be the number of elements in the jth run which are larger than τ_i, plus the number of elements in the $j+1$st run which are smaller than τ_i. For example, if $\tau = 385924617$, then the values of $w_1, w_2, \ldots, w_9$ are $1, 1, 3, 3, 3, 2, 1, 2, 1$.

THEOREM 5.3. *Given* $\tau \in S_n$,

$$(5.3) \qquad \sum_{P \in cars(\tau)} q^{dinv(P)} t^{area(P)} = t^{maj(\tau)} \prod_{i=1}^{n} [w_i(\tau)]_q.$$

PROOF. We will build up elements of cars(τ) by starting at the end of τ, where elements go in rows of length 0, and adding elements right to left. We define a partial parking function to be a Dyck path $\pi \in L_{m,m}^{+}$ for some m, together with a placement of m distinct positive integers (not necessarily the integers 1 through m) to the right of the N steps of π, with strict decrease down columns. Say $\tau = 385924617$ and we have just added car 9 to obtain a partial parking function A with cars 1 and 7 in rows of length 0, cars 2, 4 and 6 in rows of length 1, and car 9 in a row of length 2, as in the upper left grid of Figure 5. The rows with $*$'s to the right are rows above which we can insert a row of length 2 with car 5 in it and still have a partial parking function. Note the number of starred rows equals $w_3(\tau)$, and that in general $w_i(\tau)$ can be defined as the number of ways to insert a row containing car τ_i into a partial parking function containing cars $\tau_{i+1}, \ldots, \tau_n$, in rows of the appropriate length, and still obtain a partial parking function.

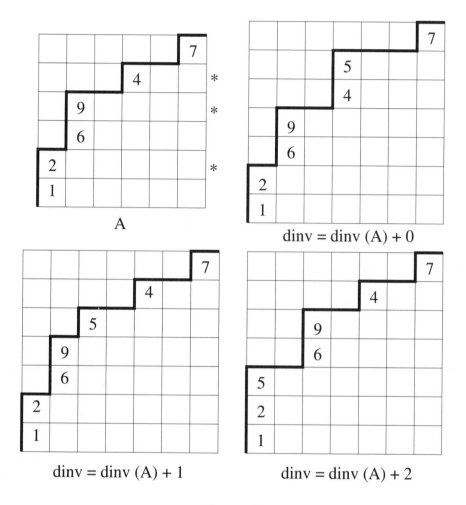

FIGURE 5

Consider what happens to dinv as we insert the row with car 5 above a starred row to form A'. Pairs of rows which form inversions in A will also form inversions in A'. Furthermore the rows of length 0 in A, or of length 1 with a car larger than 5, cannot form inversions with car 5 no matter where it is inserted. However, a starred row will form an inversion with car 5 if and only if car 5 is in a row below it. It follows that if we weight insertions by q^{dinv}, inserting car 5 at the various places will generate a factor of $[w_i(\tau)]$ times the weight of A, as in Figure 5. Finally note that for any path π corresponding to an element of cars(τ), maj$(\tau) = $ area(π). $\square$

By summing Theorem 5.3 over all $\tau \in S_n$ we get the following.

COROLLARY 5.3.1.

$$(5.4) \qquad \sum_{P \in \mathcal{P}_n} q^{dinv(P)} t^{area(P)} = \sum_{\tau \in S_n} t^{maj(\tau)} \prod_{i=1}^{n} [w_i(\tau)]_q.$$

OPEN PROBLEM 5.4. *Prove*

$$(5.5) \qquad \sum_{P \in \mathcal{P}_n} q^{dinv(P)} t^{area(P)}$$

is symmetric in q, t.

REMARK 5.5. Beyond merely proving the symmetry of (5.5), one could hope to find a bijective proof. It is interesting to note that by (5.3) the symmetry in q, t when one variable equals 0 reduces to the fact that both inv and maj are Mahonian. Hence any bijective proof of symmetry may have to involve generalizing Foata's bijective transformation of maj into inv.

The Statistic area$'$

By a *diagonal labeling* of a Dyck path $\pi \in L_{n,n}^+$ we mean a placement of the numbers 1 through n in the squares on the main diagonal $y = x$ in such a way that for every consecutive EN pair of steps of π, the number in the same column as the E step is smaller than the number in the same row as the N step. Let $\mathcal{A}_n$ denote the set of pairs (A, π) where A is a diagonal labeling of $\pi \in L_{n,n}^+$. Given such a pair (A, π), we let area$'(A, \pi)$ denote the number of area squares x of π for which the number on the diagonal in the same column as x is smaller than the number in the same row as x. Also define bounce$(A, \pi) =$ bounce(π).

The following result appears in [**HL05**].

THEOREM 5.6. *There is a bijection between* $\mathcal{P}_n$ *and* $\mathcal{A}_n$ *which sends* (dinv, area) *to* (area$'$, bounce).

PROOF. Given $P \in \mathcal{P}_n$ with associated path π, we begin to construct a pair $(A, \zeta(\pi))$ by first letting $\zeta(\pi)$ be the same path formed by the ζ map from the proof of Theorem 3.15. The length α_1 of the first bounce of ζ is the number of rows of π of length 0, etc.. Next place the cars which occur in P in the rows of length 0 in the lowest α_1 diagonal squares of ζ, in such a way that the order in which they occur, reading top to bottom, in P is the same as the order in which they occur, reading top to bottom, in ζ. Then place the cars which occur in P in the rows of length 1 in the next α_2 diagonal squares of ζ, in such a way that the order in which they occur, reading top to bottom, in P is the same as the order in which they occur, reading top to bottom, in ζ. Continue in this way until all the diagonal squares are filled, resulting in the pair (A, ζ). See Figure 6 for an example.

The properties of the ζ map immediately imply area$(\pi) =$ bounce(A, ζ) and that $(A, \zeta) \in \mathcal{A}_n$. The reader will have no trouble showing that the equation dinv$(P) =$ area$'(A, \zeta)$ is also implicit in the proof of Theorem 3.15. $\qquad\square$

The pmaj Statistic

We now define a statistic on parking functions called pmaj, due to Loehr and Remmel [**LR04**], [**Loe05a**] which generalizes the bounce statistic. Given $P \in \mathcal{P}_n$, we define the pmaj-parking order, denoted $\beta(P)$, by the following procedure. Let

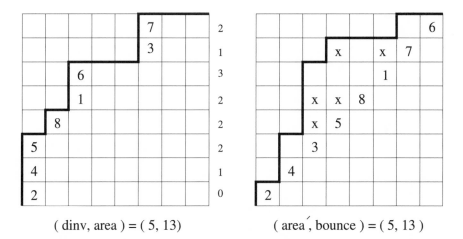

(dinv, area) = (5, 13)   (area´, bounce) = (5, 13)

FIGURE 6. The map in the proof of Theorem 5.6. Squares contributing to area′ are marked with x's.

$C_i = C_i(P)$ denote the set of cars in column i of P, and let β_1 be the largest car in C_1. We begin by parking car β_1 in spot 1. Next we perform the "dragnet" operation, which takes all the cars in $C_1/\{\beta_1\}$ and combines them with C_2 to form C_2'. Let β_2 be the largest car in C_2' which is smaller then β_1. If there is no such car, let β_2 be the largest car in C_2'. Park car β_2 in spot 2 and then iterate this procedure. Assuming we have just parked car β_{i-1} in spot $i-1$, $3 \le i < n$, we let $C_i' = C_{i-1}'/\{\beta_{i-1}\}$ and let β_i be the largest car in C_i' which is smaller than β_{i-1}, if any, while otherwise β_i is the largest car in C_i'. For the example in Figure 7, we have $C_1 = \{5\}$, $C_2 = \{1, 7\}$, $C_3 = \{\}$, etc. and $C_2' = \{1, 7\}$, $C_3' = \{7\}$, $C_4' = \{2, 4, 6\}$, $C_5' = \{2, 3, 4\}$, etc., with $\beta = 51764328$.

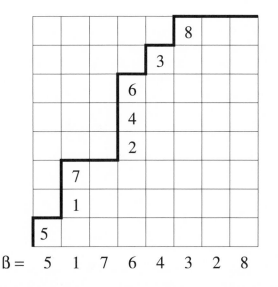

$\beta =$ 5 1 7 6 4 3 2 8

FIGURE 7. A parking function P with pmaj parking order $\beta = 51764328$.

Now let $\text{rev}(\beta(P)) = (\beta_n, \beta_{n-1}, \ldots, \beta_1)$ and define $\text{pmaj}(P) = \text{maj}(\text{rev}(\beta(P)))$. For the parking function of Figure 7 we have $\text{rev}(\beta) = 82346715$ and $\text{pmaj} = 1+6 = 7$. Given $\pi \in L_{n,n}^+$, it is easy to see that if P is the parking function for π obtained by placing car i in row i for $1 \le i \le n$, then $\text{pmaj}(P) = \text{bounce}(\pi)$. We call this parking function the primary pmaj parking function for π.

We now describe a bijection $\Gamma : \mathcal{P}_n \to \mathcal{P}_n$ from [**LR04**] which sends $(\text{area}, \text{pmaj})$ $\to (\text{dinv}, \text{area})$. The crucial observation behind it is this. Fix $\gamma \in S_n$ and consider the set of parking functions which satisfy $\text{rev}(\beta(P)) = \gamma$. We can build up this set recursively by first forming a partial parking function consisting of car γ_n in column 1. If $\gamma_{n-1} < \gamma_n$, then we can form a partial parking function consisting of two cars whose pmaj parking order is $\gamma_n \gamma_{n-1}$ in two ways. We can either have both cars γ_n and γ_{n-1} in column 1, or car γ_n in column 1 and car γ_{n-1} in column 2. If $\gamma_{n-1} > \gamma_n$, then we must have car γ_n in column 1 and car γ_{n-1} in column 2. In the case where $\gamma_{n-1} < \gamma_n$, there were two choices for columns to insert car γ_{n-1} into, corresponding to the fact that $w_{n-1}(\gamma) = 2$. When $\gamma_{n-1} > \gamma_n$, there was only one choice for the column to insert γ_{n-1} into, and correspondingly $w_{n-1}(\gamma) = 1$.

More generally, say we have a partial parking function consisting of cars in the set $\{\gamma_n, \ldots, \gamma_{i+1}\}$ whose pmaj parking order is $\gamma_n \cdots \gamma_{i+2}\gamma_{i+1}$. It is easy to see that the number of ways to insert car γ_i into this so the new partial parking function has pmaj parking order $\gamma_n \cdots \gamma_{i+1}\gamma_i$ is exactly $w_i(\gamma)$. Furthermore, as you insert car γ_i into columns $n-i+1, n-i, \ldots, n-i-w_i(\gamma)+2$ the area of the partial parking functions increases by 1 each time. It follows that

$$(5.6) \qquad \sum_{P \in \mathcal{P}_n} q^{\text{area}(P)} t^{\text{pmaj}(P)} = \sum_{\gamma \in S_n} t^{\text{maj}(\gamma)} \prod_{i=1}^{n-1} [w_i(\gamma)].$$

Moreover, we can identify the values of $(\text{area}, \text{pmaj})$ for individual parking functions by considering permutations $\gamma \in S_n$ and corresponding n-tuples $(u_1, \ldots, u_n)$ with $0 \le u_i < w_i(\gamma)$ for $1 \le i \le n$. (Note u_n always equals 0). Then $\text{maj}(\gamma) = \text{pmaj}(P)$, and $u_1 + \ldots + u_n = \text{area}(P)$. Expressed in terms of preference functions, we have $f(\beta_{n+1-i}) = n+1-i-u_i$ for $1 \le i \le n$.

Now given such a pair $\gamma \in S_n$ and corresponding n-tuple $(u_1, \ldots, u_n)$, from the proof of Theorem 5.3 we can build up a parking function Q recursively by inserting cars $\gamma_n, \gamma_{n-1}, \ldots$ one at a time, where for each j the insertion of car γ_j adds u_j to $\text{dinv}(Q)$. Thus we end up with a bijection $\Gamma : P \sim Q$ with $(\text{area}(P), \text{pmaj}(P)) = (\text{dinv}(Q), \text{area}(Q))$. The top of Figure 8 gives the various partial parking functions in the construction of P, and after those are the various partial parking functions in the construction of Q, for $\gamma = 563412$ and $(u_1, \ldots, u_6) = (2, 0, 1, 0, 1, 0)$.

COROLLARY 5.6.1. *The marginal distributions of pmaj, area, and dinv over $\mathcal{P}_n$ are all the same, i.e.*

$$(5.7) \qquad \sum_{P \in \mathcal{P}_n} q^{\text{pmaj}(P)} = \sum_{P \in \mathcal{P}_n} q^{\text{area}(P)} = \sum_{P \in \mathcal{P}_n} q^{\text{dinv}(P)}.$$

EXERCISE 5.7. Notice that in Figure 8, the final $\Gamma : P \to Q$ correspondence is between parking functions which equal the primary pmaj and Maxdinv parking functions for their respective paths. Show that this is true in general, i.e. that when P equals the primary pmaj parking function for π then the bijection Γ reduces to the inverse of the bijection ζ from the proof of Theorem 3.15.

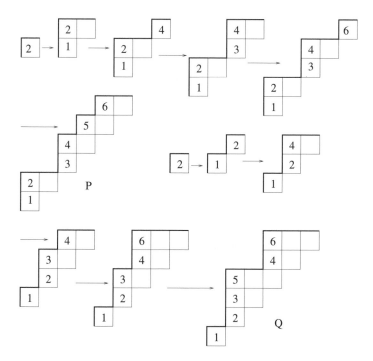

FIGURE 8. The recursive construction of the P and Q parking functions in the Γ correspondence for $\gamma = 563412$ and $u = (2, 0, 1, 0, 1, 0)$.

The Cyclic-Shift Operation

Given $S \subseteq \{1, \ldots, n\}$, let $\mathcal{P}_{n,S}$ denote the set of parking functions for which $C_1(P) = S$. If x is a positive integer, define

$$(5.8) \qquad CYC_n(x) = \begin{cases} x + 1 & \text{if } x < n \\ 1 & \text{if } x = n. \end{cases}$$

For any set S of positive integers, let $CYC_n(S) = \{CYC_n(x) : x \in S\}$. Assume $S = \{s_1 < s_2 < \cdots < s_k\}$ with $s_k < n$. Given $P \in \mathcal{P}_{n,S}$, define the cyclic-shift of P, denoted $CYC_n(P)$, to be the parking function obtained by replacing C_i, the cars in column i of P, with $CYC_n(C_i)$, for each $1 \le i \le n$. Note that the column of P containing car n will have to be sorted, with car 1 moved to the bottom of the column. The map $CYC_n(P)$ is undefined if car n is in column 1. See the top portion of Figure 9 for an example.

PROPOSITION 5.7.1. [**Loe05a**] *Suppose $P \in \mathcal{P}_{n,S}$ with $S = \{s_1 < s_2 < \cdots < s_k\}$, $s_k < n$. Then*

$$(5.9) \qquad pmaj(P) = pmaj(CYC(P)) + 1.$$

PROOF. Imagine adding a second coordinate to each car, with car i initially represented by (i, i). If we list the second coordinates of each car as they occur in the pmaj parking order for P, by definition we get the sequence $\beta_1(P)\beta_2(P) \cdots \beta_n(P)$. We now perform the cyclic-shift operation to obtain $CYC(P)$, but when doing so

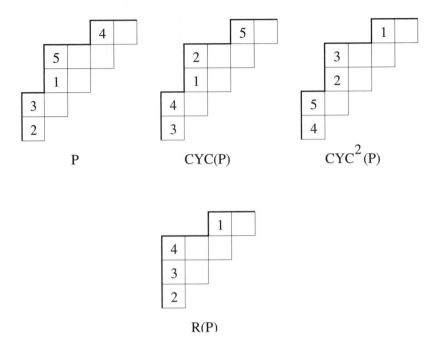

FIGURE 9. The map $R(P)$.

we operate only on the first coordinates of each car, leaving the second coordinates unchanged. The reader will have no trouble verifying that if we now list the second coordinates of each car as they occur in the pmaj parking order for $CYC(P)$, we again get the sequence $\beta_1(P)\beta_2(P)\cdots\beta_n(P)$. It follows that the pmaj parking order of $CYC(P)$ can be obtained by starting with the pmaj parking order β of P and performing the cyclic-shift operation on each element of β individually. See Figure 10.

Say n occurs in the permutation $\operatorname{rev}(\beta(P))$ in spot j. Note that we must have $j < n$, or otherwise car n would be in column 1 of P. Clearly when we perform the cyclic-shift operation on the individual elements of the permutation $\operatorname{rev}(\beta(P))$ the descent set will remain the same, except that the descent at j is now replaced by a descent at $j-1$ if $j > 1$, or is removed if $j = 1$. In any case the value of the major index of $\operatorname{rev}(\beta(P))$ is decreased by 1. □

Using Proposition 5.7.1, Loehr derives the following recurrence.

THEOREM 5.8. *Let $n \geq 1$ and $S = \{s_1 < \cdots < s_k\} \subseteq \{1,\ldots,n\}$. Set*

$$(5.10) \qquad P_{n,S}(q,t) = \sum_{P \in \mathcal{P}_{n,S}} q^{area(P)} t^{pmaj(P)}.$$

Then

$$(5.11) \qquad P_{n,S}(q,t) = q^{k-1} t^{n-s_k} \sum_{T \subseteq \{1,\ldots,n\}/S} P_{n-1,CYC_n^{n-s_k}(S \cup T/\{s_k\})}(q,t),$$

with the initial conditions $P_{n,\emptyset}(q,t) = 0$ for all n and $P_{1,\{1\}}(q,t) = 1$.

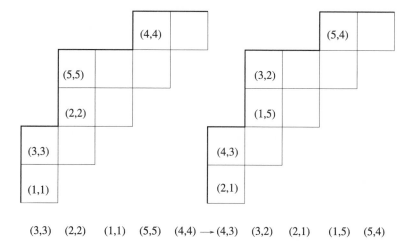

$$(3,3) \quad (2,2) \quad (1,1) \quad (5,5) \quad (4,4) \longrightarrow (4,3) \quad (3,2) \quad (2,1) \quad (1,5) \quad (5,4)$$

FIGURE 10. The cyclic-shift operation and the pmaj parking order.

PROOF. Let $Q = CYC_n^{n-s_k}(P)$. Then

$$(5.12) \qquad\qquad \mathrm{pmaj}(Q) + n - s_k = \mathrm{pmaj}(P)$$
$$\mathrm{area}(Q) = \mathrm{area}(P).$$

Since car n is in the first column of Q, in the pmaj parking order for Q car n is in spot 1. By definition, the dragnet operation will then combine the remaining cars in column 1 of Q with the cars in column 2 of Q. Now car n being in spot 1 translates into car n being at the end of $\mathrm{rev}(\beta(Q))$, which means $n-1$ is not in the descent set of $\mathrm{rev}(\beta(Q))$. Thus if we define $R(P)$ to be the element of $\mathcal{P}_{n-1}$ obtained by parking car n in spot 1, performing the dragnet, then truncating column 1 and spot 1 as in Figure 9, we have

$$(5.13) \qquad\qquad \mathrm{pmaj}(R(P)) = \mathrm{pmaj}(P) - (n - s_k).$$

Furthermore, performing the dragnet leaves the number of area cells in columns $2, 3, \ldots, n$ of Q unchanged but eliminates the $k-1$ area cells in column 1 of Q. Thus

$$(5.14) \qquad\qquad \mathrm{area}(R(P)) = \mathrm{area}(P) - k + 1$$

and the recursion now follows easily. $\qquad\qquad\qquad\qquad\qquad\qquad\qquad\square$

Loehr also derives the following compact formula for $P_{n,S}$ when $t = 1/q$.

THEOREM 5.9. *For $n \geq 0$ and $S = \{s_1 < \cdots < s_k\} \subseteq \{1, \ldots, n\}$,*

$$(5.15) \qquad\qquad q^{\binom{n}{2}} P_{n,S}(1/q, q) = q^{n-k}[n]^{n-k-1} \sum_{x \in S} q^{n-x}.$$

PROOF. Our proof is, for the most part, taken from [**Loe05a**]. If $k = n$,

$$(5.16) \qquad\qquad P_{n,S}(1/q, q) = P_{n,\{1,\ldots,n\}}(1/q, q) = q^{-\binom{n}{2}},$$

while the right-hand-side of (5.15) equals $[n]^{-1}[n] = 1$. Thus (5.15) holds for $k = n$. It also holds trivially for $n = 0$ and $n = 1$. So assume $n > 1$ and $0 < k < n$. From (5.11),

$$(5.17) \quad P_{n,S}(1/q, q) = q^{n+1-s_k-k} \sum_{T \subseteq \{1,\dots,n\}/S} P_{n-1, C_n^{n-s_k}(S \cup T/\{s_k\})}(1/q, q)$$

$$= q^{n+1-s_k-k} \sum_{j=0}^{n-k} \sum_{\substack{T \subseteq \{1,\dots,n\}/S \\ |T|=j}} P_{n-1, C_n^{n-s_k}(S \cup T/\{s_k\})}(1/q, q).$$

The summand when $j = n - k$ equals

$$(5.18) \quad P_{n-1, \{1,\dots,n-1\}}(1/q, q) = q^{-\binom{n-1}{2}}.$$

For $0 \le j < n - k$, by induction the summand equals

$$(5.19) \quad q^{n-1-(j+k-1)-\binom{n-1}{2}}[n-1]^{n-1-(j+k-1)-1} \sum_{x \in C_n^{n-s_k}(S \cup T/\{s_k\})} q^{n-1-x},$$

since $j + k - 1 = |C_n^{n-s_k}(S \cup T/\{s_k\})|$. Plugging (5.19) into (5.17) and reversing summation we now have

$$(5.20) \quad q^{-2n+s_k+k}q^{\binom{n}{2}}P_{n,S}(1/q, q) = 1 + \sum_{j=0}^{n-k-1} q^{n-k-j}[n-1]^{n-k-j-1} \sum_{x=1}^{n-1} q^{n-1-x}$$

$$\times \sum_T \chi(T \subseteq \{1,\dots,n\}/S, |T| = j, C_n^{s_k-n}(x) \in S \cup T/\{s_k\}).$$

To compute the inner sum over T above, we consider two cases.

(1) $x = n - (s_k - s_i)$ for some $i \le k$. Since $x < n$, this implies $i < k$, and since $C_n^{s_k-n}(x) = s_i$, we have $C_n^{s_k-n}(x) \in S \cup T/\{s_k\}$. Thus the inner sum above equals the number of j-element subsets of $\{1,\dots,n\}/S$, or $\binom{n-k}{j}$.

(2) $x \ne n - (s_k - s_i)$ for all $i \le k$. By Exercise 5.10, the inner sum over T in (5.20) equals $\binom{n-k-1}{j-1}$.

Applying the above analysis to (5.20) we now have

(5.21)

$$q^{-2n+s_k+k}q^{\binom{n}{2}}P_{n,S}(1/q, q) = 1 + \sum_{j=0}^{n-k-1} q^{n-k-j}[n-1]^{n-k-j-1}$$

$$\times \sum_{x \text{ satisfies (1)}} \left[\binom{n-k-1}{j} + \binom{n-k-1}{j-1}\right] q^{n-1-x}$$

$$+ \sum_{j=0}^{n-k-1} q^{n-k-j}[n-1]^{n-k-j-1} \sum_{x \text{ satisfies (2)}} \binom{n-k-1}{j-1} q^{n-1-x}.$$

Now x satisfies (1) if and only if $n - 1 - x = s_k - s_i - 1$ for some $i < k$, and so

(5.22)

$$q^{-2n+s_k+k}q^{\binom{n}{2}}P_{n,S}(1/q,q) = 1 + \sum_{j=0}^{n-k-1} q^{n-k-j}[n-1]^{n-k-j}\binom{n-k-1}{j-1}$$

$$+ \sum_{j=0}^{n-k-1} q^{n-k-j-1}[n-1]^{n-k-j-1}\binom{n-k-1}{j}\sum_{i=1}^{k-1}q^{s_k-s_i}$$

$$= \sum_{\substack{m=0 \\ (m=j-1)}}^{n-k-1} \binom{n-k-1}{m}(q[n-1])^{n-k-m-1}$$

$$+ \sum_{i=1}^{k-1}q^{s_k-s_i}\sum_{j=0}^{n-k-1}(q[n-1])^{n-k-j-1}\binom{n-k-1}{j}$$

$$= (1+q[n-1])^{n-k-1} + \sum_{i=1}^{k-1}q^{s_k-s_i}(1+q[n-1])^{n-k-1}$$

$$= [n]^{n-k-1}(1+\sum_{i=1}^{k-1}q^{s_k-s_i}).$$

Thus

(5.23) $$q^{\binom{n}{2}}P_{n,S}(1/q,q) = q^{n-k}[n]^{n-k-1}\sum_{x\in S}q^{s_k-x+n-s_k}.$$

□

As a corollary of Theorem 2.14, Haiman proves a conjecture he attributes in [**Hai94**] to Stanley, namely that

(5.24) $$q^{\binom{n}{2}}\mathcal{H}(DH_n;1/q,q) = [n+1]^{n-1}.$$

Theorem 5.15 and (5.24) together imply Conjecture 5.2 is true when $t = q$ and $q = 1/q$. To see why, first observe that

(5.25) $$P_{n+1,\{n+1\}}(q,t) = \sum_{P\in\mathcal{P}_n}q^{\mathrm{area}(P)}t^{\mathrm{pmaj}(P)}.$$

Hence by Theorem 5.15,

(5.26) $$q^{\binom{n}{2}}\sum_{P\in\mathcal{P}_n}q^{-\mathrm{area}(P)}q^{\mathrm{pmaj}(P)} = q^{\binom{n}{2}}P_{n+1,\{n+1\}}(1/q,q)$$

$$= q^{-n}q^{\binom{n+1}{2}}P_{n+1,\{n+1\}}(1/q,q)$$

$$= q^{-n}q^n[n+1]^{n-1}q^0 = [n+1]^{n-1}.$$

EXERCISE 5.10. Show that if $x \neq n - (s_k - s_i)$ for all $i \leq k$, the inner sum over T in (5.20) equals $\binom{n-k-1}{j-1}$.

The main impediment to proving Conjecture 5.2 seems to be the lack of a recursive decomposition of the Hilbert series which can be expressed using the ∇

operator. As a partial result in this direction, Loehr [**Loe05a**] conjectures that

$$(5.27) \qquad P_{n,\{1,2,\dots,k\}}(q,t) = q^{\binom{k}{2}} t^{n-k} \left\langle \nabla e_{n-k} \left[X \frac{1-q^k}{1-q} \right], h_{1^{n-k}} \right\rangle,$$

and poses the following problem.

OPEN PROBLEM 5.11. *Find an expression for $P_{n,S}(q,t)$ in terms of the ∇ operator for general S.*

EXERCISE 5.12. A special case of Conjecture 6.34 discussed in the next chapter predicts that

$$(5.28) \qquad \langle \nabla E_{n,k}, h_{1^n} \rangle = \sum_{\substack{P \in \mathcal{P}_n \\ \pi(P) \text{ has } k \text{ rows of length } 0}} q^{\operatorname{dinv}(P)} t^{\operatorname{area}(P)}.$$

Show (5.28) implies (5.27).

Hint: Use the Γ bijection and (3.24) to translate (5.27) into a statement involving dinv and the $\nabla E_{n,k}$.

CHAPTER 6

The Shuffle Conjecture

A Combinatorial Formula for the Character of the Space of Diagonal Harmonics

Recall (Proposition 2.17.1) that $\mathcal{F}(DH_n; q, t) = \nabla e_n$. In this section we discuss a recent conjecture of Haglund, Haiman, Loehr, Remmel and Ulyanov [**HHL$^+$05c**] which gives a combinatorial description of the expansion of ∇e_n into monomials. It contains the conjecture for the Hilbert series of DH_n from Chapter 5 and the formula for the q, t-Schröder formula from Chapter 4 as special cases.

Given a Dyck path $\pi \in L_{n,n}^+$, let a *word parking function* σ for π be a placement of positive integers k, $k \in \{1, 2, \ldots, n\}$, in the parking squares of π, with strict decrease down columns. Thus, a word parking function is a parking function with possibly repeated cars. Furthermore, extend the definition of dinv to word parking functions in the natural way by counting the number of pairs of cars in the same diagonal, with the larger car above, or in successive diagonals, with the larger car below and left in the sense of Definition 3.14. Note that equal cars never contribute anything to dinv. The reading word read(σ) of a word parking function is defined by reading along diagonals in a SE direction, outside to in, as in the case of parking functions, and area(σ) is the area of the underlying Dyck path. Let $\mathcal{WP}_{n,\pi}$ denote the set of word parking functions for a path π, and let $\mathcal{WP}_n$ denote the union of these sets over all $\pi \in L_{n,n}^+$. Set x^σ equal to the product, over all cars $c \in \sigma$, of x^c, i.e. x_1 to the number of times car 1 appears times x_2 to the number of times car 2 appears, etc. See Figure 1 for an example.

CONJECTURE 6.1 (HHLRU05).

$$(6.1) \qquad \nabla e_n = \sum_{\sigma \in \mathcal{WP}_n} x^\sigma q^{dinv(\sigma)} t^{area(\sigma)}.$$

Given an ordered sequence of distinct positive integers $r_1, r_2, \ldots, r_m$ with $1 \leq r_i \leq n$, we say $\sigma \in S_n$ is a $r_1, \ldots, r_m$-shuffle if for each i, $1 \leq i < m$, r_i occurs before r_{i+1} in $\sigma_1 \sigma_2 \cdots \sigma_n$. Given a composition μ of n, we say $\sigma \in S_n$ is a μ-shuffle if it is a shuffle of each of the increasing sequences $[1, 2, \ldots, \mu_1]$, $[\mu_1 + 1, \ldots, \mu_1 + \mu_2]$, $\ldots$ of lengths $\mu_1, \mu_2, \ldots$. For example, σ is a $2, 3, 2$ shuffle if it is a shuffle of $[1, 2]$, $[3, 4, 5]$, and $[6, 7]$. The 6 permutations in S_4 which are $(2), (2)$ shuffles are

(6.2) $\qquad$ 1234 $\quad$ 1324 $\quad$ 1342 $\quad$ 3124 $\quad$ 3142 $\quad$ 3412.

Start with a word parking function σ, for a path π, with μ_1-1's, μ_2-2's, etc. We construct the "standardization" σ' of σ by replacing all the 1's in σ by the numbers $1, 2, \ldots, \mu_1$, all the 2's by the numbers $\mu_1 + 1, \ldots, \mu_1 + \mu_2$, etc., in such a way that the reading word of σ' is a μ-shuffle. A moment's thought shows there is a unique way to do this. One easily checks that dinv(σ) = dinv(σ'), which shows

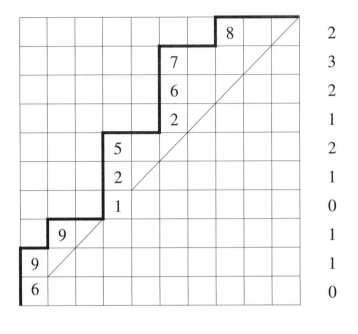

$$q^6 t^{13} x_1 x_2^2 x_5 x_6^2 x_7 x_8 x_9^2$$

FIGURE 1. A word parking function with dinv $= 6$.

that Conjecture 6.1 can be rephrased as the statement that

$$(6.3) \qquad \nabla e_n|_{m_\mu} = \sum_{\substack{P \in \mathcal{P}_n \\ P \text{ a } \mu\text{-shuffle}}} q^{\text{dinv}(P)} t^{\text{area}(P)}$$

$$(6.4) \qquad\qquad\qquad = \langle \nabla e_n, h_\mu \rangle .$$

Since $\langle \nabla e_n, h_{1^n} \rangle = \mathcal{H}_n(DH_n; q, t)$, and since the set of parking functions which are 1^n-shuffles is the set of all parking functions, Conjecture 6.1 contains Conjecture 5.2 as a special case.

Path Symmetric Functions and LLT Polynomials

In 1997 Lascoux, Leclerc and Thibon [**LLT97**], [**LT00**] introduced a new family of symmetric functions now known as LLT polynomials. They were originally defined as a sum, over objects known as ribbon tableaux (described below) of some skew shape μ, of monomials in X times a power of q, the "cospin" of the ribbon tableau. It is not at all obvious from their definition that they are symmetric functions, a fact that Lascoux, Leclerc and Thibon proved by an algebraic argument. Using the "quotient map" of Littlewood, it is known that ribbon tableaux T of a given skew shape μ are in bijection with n-tuples $\tilde{T}$ of SSYT of various skew shapes. Schilling, Shimizono and White [**SSW03**] showed that the cospin of a given ribbon tableau T could be expressed as the number of "inversion triples" of the corresponding tuple $\tilde{T}$ of SSYT. Their definition of inversion triples is somewhat complicated to describe, but Haiman and his student Michelle Bylund independently found that cospin(T) equals the number of "inversion pairs" of $\tilde{T}$, minus a constant depending on μ but not on T. Their definition of inversion pairs is based on the dinv statistic for parking functions, and is easier to work with than cospin or the statistic of Schilling, Shimizono and White. Bylund and Haiman were able to use their statistic to give the first purely combinatorial proof of the symmetry of LLT polynomials; their proof forms the appendix of [**HHL05a**].

An n-ribbon R is a connected skew shape of n squares with no subshape of 2×2 squares. The content of a given square with (column, row) coordinates (i, j) in an n-ribbon is $j - i$. The square in R of maximum content, i.e. the rightmost, lowest square, is called the *tail*. The content of a ribbon is the content of its tail. A ribbon tableau T of shape μ is a filling of a skew shape μ by n-ribbons (for some fixed n), and a filling σ of the squares of the n-ribbons by positive integers, such that

(a) σ is constant within any given ribbon,
(b) there is weak increase across rows and up columns, and
(c) if the tail of a ribbon R sits atop another ribbon R', then the number in the tail of R' is greater than the number in the tail of R.

The spin of a ribbon is one less than the number of its rows. Let sp(T) denote the sum of the spins of all the ribbons in T, and let spmax(μ) be the maximum value of sp(T) over all ribbon tableau of shape μ. Define the statistic cospin(T) = (spmax(μ) − sp(T))/2. For any given skew shape μ, define

$$(6.5) \qquad LLT_\mu(X; q) = \sum_T x^T q^{\mathrm{cospin}(T)},$$

where the sum is over all ribbon tableau of μ, and $x^T = x_1^{\#1\text{'s in } T} x_2^{\#2\text{'s in } T} \cdots$.

EXERCISE 6.2. Let $|\mu| = 2n$. Show that if μ can be tiled by two n-ribbons whose cells have at least one content in common, then there are exactly two tilings T, T' of μ, and $sp(T)' = sp(T) + 1$. Here we define $S(\theta)$ for a ribbon θ as one less than the number of rows of θ, and $S(T)$ as the sum of $S(\theta)$ over all $\theta \in T$, and finally

$$(6.6) \qquad sp(T) = \frac{1}{2}(S(T) - \mathrm{smin}(\mu)),$$

where smin(μ) is the minimum value of $S(T)$ over all tilings T of μ.

THEOREM 6.3. *(Lascoux, Leclerc, Thibon) For any skew shape μ, $LLT_\mu(X; q)$ is a symmetric function.*

In the quotient map which takes a ribbon tableau T to a tuple $\tilde{T}$ of SSYT, each ribbon θ of T gets sent to a single square $\tilde{\theta}$ of $\tilde{T}$ (containing the same integer), and if the content of the tail of θ is of the form $rn + p$, where $0 \le p < n$, then the content of $\tilde{\theta} = r$, and furthermore $\tilde{\theta}$ is in the $p + 1$st element of the tuple $\tilde{T}$. In addition, if T' is another ribbon tableau of shape ν with $\mu \subset \nu$, and $T \subset T'$, then for each k, $1 \le k \le n$, the kth skew shape in $\tilde{T}'$ contains the kth skew shape in $\tilde{T}$. See Figure 2.

To describe Bylund and Haiman's inversion statistic, we begin with a tuple $\tilde{T} = (\tilde{T}^{(1)}, \ldots, \tilde{T}^{(n)})$ of n SSYT, and we arrange the skew-shapes in the tuple geometrically so that all the squares with content 0 end up on the same "master

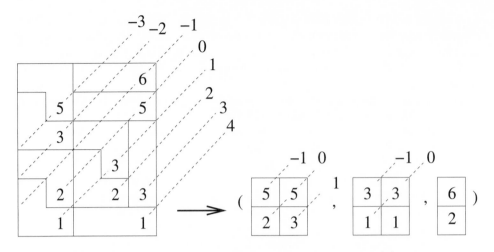

FIGURE 2. On the left, a ribbon tableau T with $\mathrm{sp}(T) = 8$, $n = 3$ and $\mu = 5^6$. The contents of squares on a given diagonal are listed above the dotted lines. On the right, the three SSYT $(\tilde{T}^{(1)}, \tilde{T}^{(2)}, \tilde{T}^{(3)})$ which correspond to T under the quotient map. Note that tails on the same diagonal on the left get mapped to squares on the same diagonal on the right.

diagonal", and similarly for squares of other content's. For example, the tuple on the right in Figure 2 becomes the shifted tuple $\tilde{T} = (\tilde{T}^{(1)}, \tilde{T}^{(2)}, \tilde{T}^{(3)})$ in Figure 3.

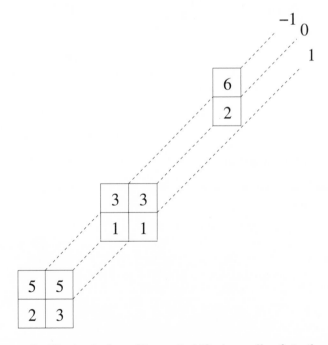

FIGURE 3. The tuple from Figure 2 shifted so cells of similar content in different tuple elements are on the same diagonal.

Now we define an inversion pair for $\tilde{T}$ to be a pair (c, d) of squares, containing numbers $\sigma(c)$ and $\sigma(d)$, respectively, with c in $\tilde{T}^{(i)}$ and d in $\tilde{T}^{(j)}$ where $1 \leq i < j \leq n$, and either

(a) $\sigma(c) > \sigma(d)$ and c and d have the same content, i.e. are on the same diagonal
  or

(b) $\sigma(c) < \sigma(d)$ and the content of c is one less than the content of d, i,e. they are on successive diagonals with d "above right".

For example, in Figure 3, the 5 in upper-right-hand corner of $\tilde{T}^{(1)}$ forms inversion pairs with the 1, 3, and 2 on the same diagonal, while the 3 in the upper-left-hand corner of $\tilde{T}^{(2)}$ forms an inversion pair with the 2 in $\tilde{T}^{(1)}$. Let $\text{inv}(\tilde{T}')$ denote the number of such inversion pairs.

THEOREM 6.4. *(Bylund, Haiman) For any skew shape μ that can be tiled by n-ribbons, there is a constant $e = e(\mu)$ such that*

$$(6.7) \qquad q^e LLT_\mu(X; q) = \sum_{\tilde{T}} q^{inv(\tilde{T})} x^{\tilde{T}},$$

where the sum is over all SSYT of shape $\tilde{\mu}$. For example, there are two tilings of the shape $(2, 2)$ by 2-ribbons, and 4 ribbon tableaux; the corresponding SSYT under the 2-quotient map are displayed in Figure 4.

REMARK 6.5. We call the polynomial defined by the right-hand-side of (6.7) an *LLT product* of $\tilde{\mu}_1, \tilde{\mu}_2, \ldots$ For a given Dyck path π, let

$$(6.8) \qquad F_\pi(X; q) = \sum_{\sigma \in WP_{n,\pi}} x^\sigma q^{\text{dinv}(\sigma)},$$

so by Conjecture 6.1, $\nabla e_n = \sum_\pi F_\pi(X; q)$. We call the F_π *path symmetric functions*. It is easy to see that $F_\pi(X; q)$ is an LLT product of vertical strips. For example, given the word parking function σ of Figure 1, by reversing the order of the columns we obtain the tuple in Figure 5. From the definitions, a pair of squares in this tuple forms a Haiman-Bylund inversion pair if and only if the pair yields a contribution to dinv in Figure 1.

Remark 6.5 shows that the $F_\pi(X; q)$ are symmetric functions. Lascoux, Leclrec and Thibon conjectured that for any skew shape μ, $LLT_\mu(X; q)$ is Schur positive. Leclerc and Thibon [**LT00**] showed that an LLT product of partition shapes is a parabolic Kazhdan-Lusztig polynomial of a certain type, and later Kashiwara and Tanisaki [**KT02**] showed these polynomials are Schur positive. In [**HHL+05c**] the authors show that this result can be extended to include LLT products where each element of the tuple is obtained by starting with a partition shape and pushing it straight up an arbitrary distance, i.e. each tuple element is of the form λ/ν, where ν is a rectangle with $\lambda_1 = \nu_1$. Since the F_π are of this form, they are hence Schur positive. Recently Grojnowski and Haiman [**GH06**] have announced a proof that any LLT polynomial is Schur positive, resolving the general conjecture. However, none of these results yield any purely combinatorial expression for the Schur coefficients.

OPEN PROBLEM 6.6. *Find a combinatorial interpretation for*

$$(6.9) \qquad\qquad \langle \mathcal{F}_\pi, s_\lambda \rangle$$

for general λ, π.

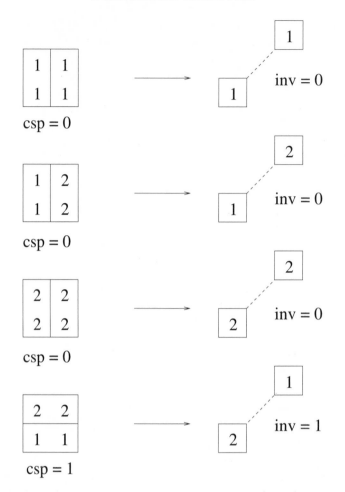

FIGURE 4. The four ribbon tableaux which generate the polynomial $LLT_{(2,2)}(X;q) = x_1^2 + x_1 x_2 + x_2^2 + q x_1 x_2$, and the corresponding SSYT of shape $((1),(1))$. Here the offset e equals 0.

Of course, Open Problem 6.6 is a special case of the more general question of finding a combinatorial interpretation for the Schur coefficients of LLT polynomials, but we list it as a separate problem anyway, as there may well be some special structure peculiar to this case which will allow it to be solved. Note that (6.4), together with the Jacobi-Trudi identity from Theorem 1.20, implies an expression for these coefficients involving parking function statistics, but with alternating signs. In another direction, combinatorial expressions are known for any LLT product of two skew-shapes (corresponding to 2-ribbons, known as domino tableaux) [**CL95**],[**vL00**]. See also [**HHL05a**]. In her Ph. D. thesis under M. Haiman, Sami Assaf [**Ass07b**] gives a combinatorial construction which proves Schur positivity for LLT products of three skew-shapes. As mentioned in the preface, very recently Assaf [**Ass07a**] has proved Schur positivity of general LLT polynomials by a 21-page, self-contained combinatorial argument. The resulting formula for the Schur coefficients she obtains involves a (possibly lengthy) inductive construction, but one

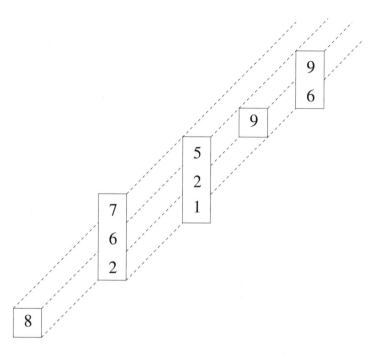

FIGURE 5

can be optimistic that further research and study of her methods will result in an elegant, closed form solution to Open Problem 6.6.

EXAMPLE 6.7. Let π, together with its 6 parking functions, be as in Figure 6. As partially explained by (3.89), we know that $\mathcal{F}_\pi(X; 1) = e_{\beta(\pi)} = e_{2,2}$, which by the Pieri rule equals $s_{2,2} + s_{2,1,1} + s_{1,1,1,1}$. Thus

$$(6.10) \qquad \mathcal{H}_\pi(q) = 2q + 3q^2 + q^3$$

$$(6.11) \qquad = f^{(2,2)}q^a + f^{(2,1,1)}q^b + f^{(1,1,1,1)}q^c$$

$$(6.12) \qquad = 2q^a + 3q^b + q^c$$

for some a, b, c. Thus we must have $a = 1, b = 2, c = 3$. These weights can be associated to the following row-strict tableau shapes, which arise when expanding $e_{2,2}$ by the Pieri rule.

$$(6.13) \qquad e_2 e_2 = s_{1^4} + s_{2,1,1} + s_{2,2}$$

$$(6.14) \qquad = \begin{matrix} 2 \\ 2 \\ 1 \\ 1 \end{matrix} \quad + \quad \begin{matrix} 2 \\ 1 \\ 1 \ \ 2 \end{matrix} \quad + \quad \begin{matrix} 1 \ \ 2 \\ 1 \ \ 2 \end{matrix}$$

By transposing these row-strict tableaux we end up with SSYT. Thus Problem 6.6 can be viewed as asking for a statistic on SSYT, depending in some way on the underlying path π, which generates $\mathcal{F}_{n,\pi}$.

The following result gives one general class of paths for which Problem 6.6 is known to have an elegant answer.

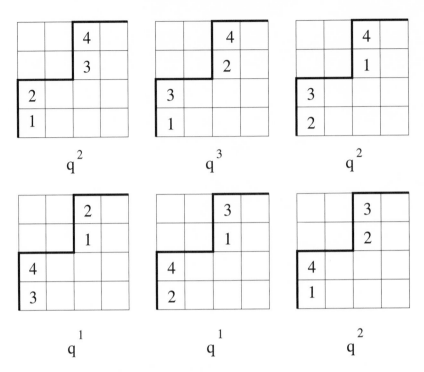

FIGURE 6. The parking functions for a balanced path.

THEOREM 6.8. *If π is a "balanced" path, i.e. π equals its own bounce path, then*

$$(6.15) \qquad \mathcal{F}_\pi = \sum_\lambda q^{mindinv(\pi)} K_{\lambda',\mu(\pi)}(q) s_\lambda,$$

where $\mu(\pi)$ is the partition obtained by rearranging the bounce steps of π into non-increasing order, mindinv is the minimum value of dinv over all $P \in \mathcal{P}_{n,\pi}$, and $K_{\beta,\alpha}(q)$ is as in (2.18).

PROOF. Once translated into a statement about Ribbon tableaux generating functions, Theorem 6.8 reduces to a result of Lascoux, Leclerc and Thibon. It can also be proved by comparing a recursive definition of $\mathcal{F}_\pi$, involving skewing operators, with similar equations for the $K_{\lambda',\mu}(q)$ obtained by Garsia and Procesi [**GP92**], as described in more detail in Section 6. □

EXAMPLE 6.9. The path in Figure 6 is balanced, with $\mu(\pi) = (2,2)$. Theorem 6.8 predicts the coefficient of $s_{2,1,1}$ equals $q^{mindinv(\pi)} K_{(3,1),(2,2)}(q)$. By the charge version of Algorithm 1.22 we have

$$(6.16) \qquad \text{charge} \begin{pmatrix} 2 & & \\ 1 & 1 & 2 \end{pmatrix} = \text{charge}(2112) = \text{comaj}(12) + \text{comaj}(21) = 1,$$

so $K_{(3,1),(2,2)}(q) = q$. Since mindinv $= 1$, the coefficient of $s_{2,1,1}$ equals q^2.

For balanced paths, mindinv equals the number of triples x, y, z of parking squares of π for which x is directly above y and z is in same diagonal as y, in a NE direction from y. For given P and such a triple, if the car in x is larger than the car in z then they contribute an inversion pair, while if it is smaller, then the car in z must perforce be larger than the car in y, so they contribute an inversion pair. On the other hand, if we let P be the parking function which generates the permutation $n \cdots 21$ when read in by columns from left to right and from top to bottom, then $\mathrm{dinv}(P)$ turns out to be precisely the number of such triples.

Superization

Given a pair of partitions μ, η, with $|\eta| + |\mu| = n$, we say $\sigma \in S_n$ is a μ, η-shuffle if σ is a shuffle of the decreasing sequences $\eta_1, \ldots, 2, 1, \eta_1 + \eta_2, \ldots, \eta_1 + 1, \ldots$, and the increasing sequences $|\eta| + 1, \ldots, |\eta| + \mu_1, |\eta| + \mu_1 + 1, \ldots, |\eta| + \mu_1 + \mu_2, \ldots$. Equivalently, σ^{-1} is the concatenation of decreasing sequences of lengths $\eta_1, \eta_2, \ldots$ followed by increasing sequences of lengths $\mu_1, \mu_2, \ldots$. This definition also applies if μ, η are any compositions.

THEOREM 6.10. *Let $f[Z]$ be a symmetric function of homogeneous degree n and $c(\sigma)$ a family of constants. If*

$$(6.17) \qquad \langle f, h_\mu \rangle = \sum_{\substack{\sigma \in S_n \\ \sigma \text{ is a } \mu\text{-shuffle}}} c(\sigma)$$

holds for all compositions μ, then

$$(6.18) \qquad \langle f, e_\eta h_\mu \rangle = \sum_{\substack{\sigma \in S_n \\ \sigma \text{ is a } \mu, \eta\text{-shuffle}}} c(\sigma)$$

holds for all pairs of compositions η, μ.

For $\pi \in L_{n,n}^+$, let $f = F_\pi[X; q]$, and for $P \in \mathcal{P}_{n,\pi}$, let $c(\mathrm{read}(P)) = q^{\mathrm{dinv}(P)}$. Then Theorem 6.10 implies that

$$(6.19) \qquad \langle F_\pi(X; q), e_\eta h_\mu \rangle = \sum_{\substack{P \in \mathcal{P}_{n,\pi} \\ \mathrm{read}(P) \text{ is a } \mu, \eta\text{-shuffle}}} q^{\mathrm{dinv}(P)}.$$

By summing this over π we get

COROLLARY 6.10.1. *(The "shuffle conjecture"). Conjecture 6.1 implies that for any compositions μ, η,*

$$(6.20) \qquad \langle \nabla e_n, e_\eta h_\mu \rangle = \sum_{\substack{P \in \mathcal{P}_n \\ \mathrm{read}(P) \text{ is a } \mu, \eta\text{-shuffle}}} q^{\mathrm{dinv}(P)} t^{\mathrm{area}(P)}.$$

OPEN PROBLEM 6.11. *By Corollary 1.46, the $\eta = \emptyset$ case of (6.20) says that the coefficient of the monomial symmetric function m_μ in ∇e_n equals the sum of $q^{\mathrm{dinv}} t^{\mathrm{area}}$ over all parking functions whose reading words are $\mu, \emptyset$-shuffles. By the Jacobi-Trudi identity (see Theorem 1.20), this implies that $\langle \nabla e_n, s_\lambda \rangle$ can be written as an alternating sum of polynomials with nonnegative integer coefficients. Refine the shuffle conjecture by finding a method to cancel the terms with negative coefficients in this identity, thus obtaining a formula for $\langle \nabla e_n, s_\lambda \rangle$ with nonnegative integer coefficients.*

Let $\mathcal{A}_\pm = \mathcal{A}_+ \cup \mathcal{A}_- = \{1 < 2 < \cdots < n < \bar{1} < \bar{2} < \cdots \bar{n}\}$ be the alphabet of positive letters $\{1, 2, \ldots, n\}$ and negative letters $\{\bar{1}, \bar{2}, \ldots, \bar{n}\}$. We let $\tilde{\sigma}$ denote a word in the alphabet $\mathcal{A}_\pm$. If $\tilde{\sigma}$ has α_1 1's, β_1 $\bar{1}$'s, α_2 2's, β_2 $\bar{2}$'s, etc., we define the standardization $\tilde{\sigma}'$ of $\tilde{\sigma}$ to be the permutation obtained by replacing the β_1 $\bar{1}$'s by the numbers $\{\beta_1, \ldots, 1\}$, so they are in *decreasing* order in $\tilde{\sigma}'$, replacing the β_2 $\bar{2}$'s by the numbers $\{\beta_1 + \beta_2, \ldots, \beta_1\}$, so they are in *decreasing* order in $\tilde{\sigma}'$, etc., then replacing the α_1 1's by the numbers $\{|\beta|+1, \ldots, |\beta|+\alpha_1\}$, so they are in *increasing* order in $\tilde{\sigma}'$, and replacing the α_2 2's by the numbers $\{|\beta|+\alpha_1+1, \ldots, |\beta|+\alpha_1+\alpha_2\}$, so they are in *increasing* order in $\tilde{\sigma}'$, etc. (thus $(\tilde{\sigma}')^{-1}$ is an α, β-shuffle). For example, the standardization of $3\bar{2}2\bar{2}31\bar{1}\bar{1}$ is 74638521. We define $\mathrm{Des}(\tilde{\sigma}) = \mathrm{Des}(\tilde{\sigma}')$.

Given two sets of variables W, Z, define the superization of a symmetric function f to be $\omega^W f[Z + W]$, where ω^W indicates ω is applied to the set of variables W only.

EXERCISE 6.12. Show that for any symmetric function f,

(6.21) $$\omega^W f[Z + W]|_{z^\mu w^\eta} = \langle f, e_\eta h_\mu \rangle.$$

For a subset D of $\{1, 2, \ldots, n-1\}$, let

(6.22) $$Q_{n,D}(Z) = \sum_{\substack{a_1 \le a_2 \le \cdots \le a_n \\ a_i = a_{i+1} \implies i \notin D}} z_{a_1} z_{a_2} \cdots z_{a_n}$$

denote Gessel's fundamental quasisymmetric function. Here $a_i \in \mathcal{A}_+$.

REMARK 6.13. For any permutation $\sigma \in S_n$, $Q_{n,\mathrm{Des}(\sigma^{-1})}(Z)$ is simply the sum of z^β, over all words β of length n in the alphabet $\mathcal{A}_+$ whose standardization equals σ.

LEMMA 6.14. *If f is a symmetric function, then (6.17) is equivalent to*

(6.23) $$f = \sum_{\sigma \in S_n} c(\sigma) Q_{n,\mathrm{Des}(\sigma^{-1})}(Z).$$

PROOF. If μ is any composition of n, by (6.22) the coefficient of $z_1^{\mu_1} \cdots z_n^{\mu_n}$ in $Q_{n,\mathrm{Des}(\sigma^{-1})}(Z)$ equals 1 if $\mathrm{Des}(\sigma^{-1}) \subseteq \{\mu_1, \mu_1 + \mu_2, \ldots\}$ and 0 otherwise. Since f is symmetric, this is equivalent to (6.17). $\square$

Proof of Theorem 6.10: Let

(6.24) $$\tilde{Q}_{n,D}(Z, W) = \sum_{\substack{a_1 \le a_2 \le \cdots \le a_n \\ a_i = a_{i+1} \in \mathcal{A}_+ \implies i \notin D \\ a_i = a_{i+1} \in \mathcal{A}_- \implies i \in D}} z_{a_1} z_{a_2} \cdots z_{a_n}$$

denote the *super quasisymmetric function*, where the indices a_i range over the alphabet $\mathcal{A}_\pm$ and for $\bar{a} \in \mathcal{A}_-$, by convention $z_{\bar{a}} = w_a$. In [**HHL**$^+$**05c**, Corollary 2.4.3] it is shown that if $f(Z) = \sum_D c_D Q_{n,D}(Z)$, where the c_D are independent of the z_i, then

(6.25) $$\omega^W f(Z + W) = \sum_D c_D \tilde{Q}_{n,D}(Z, W).$$

Thus by Lemma 6.14, (6.17) implies

(6.26) $$\omega^W f(Z + W) = \sum_{\sigma \in S_n} c(\sigma) \tilde{Q}_{n,\mathrm{Des}(\sigma^{-1})}(Z, W).$$

By (6.21), the coefficient of $z^\mu w^\eta$ in the right-hand side of (6.26) equals $\langle f, e_\eta h_\mu \rangle$. On the other hand the coefficient of $z^\mu w^\eta$ in $\tilde{Q}_{n,\mathrm{Des}(\sigma^{-1})}(Z, W)$ equals 1 if σ is a μ, η-shuffle and 0 otherwise, so (6.18) follows from (6.26).

COROLLARY 6.14.1. *For any π,*

$$(6.27) \qquad F_\pi(X;q) = \sum_S Q_{n,S}(X) \sum_{\substack{P \in \mathcal{P}_{n,\pi} \\ Des(read(P)^{-1})=S}} q^{dinv(P)}.$$

PROOF. In Lemma 6.14 let

$$(6.28) \qquad c(\sigma) = \begin{cases} q^{\mathrm{dinv}(P)} & \text{if } read(P) = \sigma \text{ and } \mathrm{Des}(read(P)^{-1}) = S \\ 0 & \text{otherwise.} \end{cases}$$

$\square$

We should also mention that for all π, $\mathcal{F}_\pi$ can be defined via parking functions whose reading words are shuffles of decreasing and increasing sequences taken in any order. For example,

$$(6.29) \qquad \langle \mathcal{F}_\pi, e_{(3,2)} h_{(5,4)} \rangle = \sum_P q^{\mathrm{dinv}(P)},$$

where the sum can be taken over all $P \in \mathcal{F}_\pi$ which are $(14, 13)$, $(8, 9, 10, 11, 12)$, $(7, 6, 5)$ and $(1, 2, 3, 4)$-shuffles.

EXERCISE 6.15. Show that the $(\mu, \eta) = (d, n - d)$ case of the shuffle conjecture is equivalent to Corollary 4.19.1, the dinv-version of Theorem 4.2.

REMARK 6.16. For any permutation $\sigma \in S_n$, $\tilde{Q}_{n,\mathrm{Des}(\sigma^1)}(Z, W)$ is the sum of z^β, over all words of length n in alphabet $\mathcal{A}_\pm$ whose standardization equals σ, where by convention we let $z_{\bar{a}} = w_a$.

The Fermionic Formula

Kerov, Kirillov and Reshetikhin [KKR86],[KR86] obtained an expression for $K_{\lambda,\mu}(q)$ as a sum of products of q-binomial coefficients times nonnegative powers of q, known as the "fermionic formula". By analogy, we refer to sums of such terms, multiplied by nonnegative powers of t, as fermionic formulas. We now derive a result which includes the fermionic formulas occurring in both Corollaries 3.10 and 5.3.1 as special cases.

Given partitions μ, η with $|\mu| + |\eta| = n$, let $N_j = \eta_1 + \ldots + \eta_j$ and $M_j = |\eta| + \mu_1 + \ldots + \mu_j$ with $N_0 = 0$, $M_0 = |\eta|$. Furthermore let $C_j = \{N_{j-1}+1, \ldots, N_j\}$, $1 \le j \le \ell(\eta)$ and let $B_j = \{M_{j-1}+1, \ldots, M_j\}$, $1 \le j \le \ell(\mu)$. Given $\sigma \in S_n$ we let $\mathrm{run}_i(\sigma)$ denote the ith run of σ, and set

$$(6.30) \qquad B_{i,j} = B_j \cap \mathrm{run}_i(\sigma)$$
$$(6.31) \qquad C_{i,j} = C_j \cap \mathrm{run}_i(\sigma).$$

We use the abbreviations

$$(6.32) \qquad b_{ij} = |B_{i,j}|, \quad c_{ij} = |C_{i,j}|.$$

Note that all of the elements of C_{ij} must occur as a consecutive block of c_{ij} elements of σ. We define $\tilde{\sigma}$ to be the permutation obtained by reversing each of these blocks of elements, leaving the remaining elements of σ fixed. Furthermore, if $\sigma_k = \min B_{ij}$ then set $V_{ij} = w_k(\sigma)$, and if $\sigma_k = \min C_{ij}$ set $W_{ij} = w_k(\sigma)$, where the w_i are as in Theorem 5.3, and where $W_{ij} = 0$ if $B_{ij} = \emptyset$ and $V_{ij} = 0$ if $C_{ij} = \emptyset$.

EXAMPLE 6.17. If $\sigma = 712346895$, $\mu = (2)$ and $\eta = (4, 3)$, we have $C_1 = \{1, 2, 3, 4\}$, $C_2 = \{5, 6, 7\}$, $B_1 = \{8, 9\}$, $\text{run}_1(\sigma) = \{7\}$, $\text{run}_2(\sigma) = \{1, 2, 3, 4, 6, 8, 9\}$, $\text{run}_3(\sigma) = \{5\}$, $B_{11} = \emptyset$, $B_{21} = \{8, 9\}$, $B_{31} = \emptyset$, $C_{11} = \emptyset$, $C_{12} = \{7\}$, $C_{21} = \{1, 2, 3, 4\}$, $C_{22} = \{6\}$, $C_{31} = \emptyset$, $C_{32} = \{5\}$, $\tilde{\sigma} = 743216895$, $W_{11} = 0$, $W_{21} = w_7(\sigma) = 2$, $V_{11} = 0$, $V_{12} = 5$, $V_{21} = 6$ and $V_{22} = 3$, $V_{31} = 0$, $V_{32} = 1$.

THEOREM 6.18. *Given $\sigma \in S_n$ such that $\tilde{\sigma}$ is a μ, η-shuffle,*

$$(6.33) \qquad \sum_{\substack{P \in cars(\sigma) \\ read(P) \text{ is a } \mu, \eta\text{-shuffle}}} q^{dinv(P)} t^{area(P)} = t^{maj(\sigma)} \prod_{i,j \geq 1} \begin{bmatrix} V_{i,j} \\ b_{ij} \end{bmatrix} \prod_{i,j \geq 1} \begin{bmatrix} W_{i,j} \\ c_{ij} \end{bmatrix} q^{\binom{c_{ij}}{2}}.$$

PROOF. Say σ, μ, η are as in Example 6.17. Consider an element $P \in cars(\sigma)$ whose reading word is a μ, η-shuffle. Then the sequence of cars in rows of length 1, read from top to bottom, must in particular be a 4321-shuffle. Thus for each such P, there will be 4! corresponding P's in $cars(\sigma)$ obtained by fixing all other cars and permuting cars $1, 2, 3, 4$ amongst themselves. Similar arguments apply to all other C_{ij} subsets and also all B_{ij} subsets, and so by Theorem 5.3 we have

$$(6.34) \qquad \sum_{\substack{P \in cars(\sigma) \\ read(P) \text{ is a } \mu, \eta\text{-shuffle}}} t^{area(P)} = t^{maj(\sigma)} \frac{\prod_{i=1}^{n-1} w_i(\sigma)}{\prod_{i,j} b_{ij}! c_{ij}!}.$$

We can insert the q-parameter into (6.34) without much more effort, since when cars $1, 2, 3, 4$ are permuted amongst themselves and other cars are left fixed, only the d-inversions involving cars $1, 2, 3, 4$ are affected. When these cars occur in decreasing order in $read(P)$, they generate $\binom{4}{2}$ inversions amongst themselves, while if they are permuted amongst themselves in all possible ways the inversions amongst themselves generate the factor $[4]!$. Similar reasoning for the increasing blocks B_{ij} leads to the result

$$(6.35) \qquad \sum_{\substack{P \in cars(\sigma) \\ read(P) \text{ is a } \mu, \eta\text{-shuffle}}} q^{dinv(P)} t^{area(P)} = t^{maj(\sigma)} \frac{\prod_{i=1}^{n-1} [w_i(\sigma)] \prod_{ij} q^{\binom{c_{ij}}{2}}}{\prod_{i,j} [b_{ij}]! [c_{ij}]!}.$$

At first glance it may seem surprising that the right-hand-side of (6.35) is a polynomial, since we are dividing by factorials, but have only q-integers in the numerator. However, note that in any string of m consecutive elements of σ, the corresponding m w-values decrease by 1 as we move left-to-right. For example, for $\sigma = 712346895$, the w-values w_2, w_3, w_4, w_5 corresponding to elements 1234 are $6, 5, 4, 3$. Thus in this case

$$(6.36) \qquad \frac{[w_2][w_3][w_4][w_5]}{[4]!} = \begin{bmatrix} 6 \\ 4 \end{bmatrix} = \begin{bmatrix} V_{21} \\ c_{21} \end{bmatrix},$$

and Theorem 6.33 is now transparent. $\qquad\qquad\qquad\qquad\qquad\qquad\qquad\qquad\square$

By summing Theorem 6.33 over relevant σ we get the following.

COROLLARY 6.18.1. *Conjecture 6.20 is equivalent to the statement that for any μ, η with $|\mu| + |\eta| = n$,*

$$(6.37) \qquad \langle \nabla e_n, e_\eta h_\mu \rangle = \sum_{\substack{\sigma \in S_n \\ \tilde{\sigma} \text{ is a } \mu, \eta\text{-shuffle}}} t^{maj(\sigma)} \prod_{i,j \geq 1} \begin{bmatrix} V_{i,j} \\ b_{ij} \end{bmatrix} \prod_{i,j \geq 1} \begin{bmatrix} W_{i,j} \\ c_{ij} \end{bmatrix} q^{\binom{c_{ij}}{2}}.$$

EXERCISE 6.19. Show that the fermionic formula for the q, t-Catalan given by Theorem 3.10 agrees with the special case $\mu = \emptyset, \eta = (n)$ of (6.37).

Skewing, Balanced Paths, and Hall-Littlewood Polynomials

Given a symmetric function $A \in \Lambda$, we define the skewing operator $A^\perp$ as the linear operator satisfying

$$\tag{6.38} \left\langle A^\perp P, Q \right\rangle = \left\langle P, AQ \right\rangle$$

for all $P, Q \in \Lambda$. Thus $A^\perp$ is adjoint to multiplication with respect to the Hall scalar product. For example, $1^\perp P = P$ for all P, and $\left\langle s_\lambda^\perp f, 1 \right\rangle$ equals the coefficient of s_λ in the Schur function expansion of f.

THEOREM 6.20. *For all λ,*

$$\tag{6.39} e_1^\perp s_\lambda = \sum_{\beta \to \lambda} s_\beta,$$

where the notation $\beta \to \lambda$ means $|\lambda/\beta| = 1$. For example, $e_1^\perp s_{3,2} = s_{2,2} + s_{3,1}$. More generally,

$$\tag{6.40} e_k^\perp s_\lambda = \sum_{\lambda/\beta \text{ is a vertical } k\text{-strip}} s_\beta$$

$$\tag{6.41} h_k^\perp s_\lambda = \sum_{\lambda/\beta \text{ is a horizontal } k\text{-strip}} s_\beta.$$

PROOF.

$$\tag{6.42} \left\langle e_k^\perp s_\lambda, s_\beta \right\rangle = \left\langle s_\lambda, e_k s_\beta \right\rangle = \begin{cases} 1 & \text{if } \lambda/\beta \text{ is a vertical } k\text{-strip} \\ 0 & \text{else} \end{cases}$$

and (6.40) follows. A similar argument applies to (6.41).   □

Note that $e_1^\perp s_2 = s_1 = e_1^\perp s_{1,1}$, which shows that a symmetric function f is not uniquely determined by $e_1^\perp f$, although we have the following simple proposition.

PROPOSITION 6.20.1. *Any homogeneous symmetric function $f \in \Lambda^{(n)}$, with $n \geq 1$, is uniquely determined by the values of $e_k^\perp f$ for $1 \leq k \leq n$, or by the values of $h_k^\perp f$ for $1 \leq k \leq n$.*

PROOF. If $e_k^\perp f = e_k^\perp g$ then for any $\gamma \vdash n - k$,

$$\tag{6.43} e_\gamma^\perp e_k^\perp f = e_\gamma^\perp e_k^\perp g.$$

Since any $\lambda \vdash n$ can be written as $\gamma \cup k$ for some k, it follows that

$$\tag{6.44} \langle f, e_\lambda \rangle = \left\langle e_\lambda^\perp f, 1 \right\rangle = \left\langle e_\lambda^\perp g, 1 \right\rangle = \langle g, e_\lambda \rangle.$$

Since the e_λ form a basis for $\Lambda^{(n)}$, we have $f = g$. The same proof works with $e_k^\perp$ replaced by $h_k^\perp$.   □

Set $\langle F_\pi(X; q), h_{1^n} \rangle = \mathcal{H}_\pi$. We can derive a recurrence for $\mathcal{H}_\pi$ by considering where car n is placed. In order to describe this recurrence, we need to introduce "generalized Dyck paths", which we define to be pairs (π, S), where π is a Dyck path and S is a subset of the parking squares of π, with the property that if $x \in S$, then any parking square of π which is in the same column as x and above x is also in S. The set of parking squares of π which are not in S is called the set of parking squares of (π, S). If $\pi \in L_{n,n}^+$, then a parking function for (π, S) is a placement of the integers 1 through $n - |S|$ in the parking squares of (π, S), with strict decrease down columns. Let $\mathcal{P}_{(\pi,S)}$ denote the set of these parking functions, and extend the definition of dinv to elements of $\mathcal{P}_{(\pi,S)}$ in the natural way by counting inversion pairs amongst pairs of cars in the same diagonal, with the larger car above, or in diagonals one apart, with the larger car below and left.

Set

$$(6.45) \qquad \mathcal{H}_{(\pi,S)}(q) = \sum_{P \in \mathcal{P}_{(\pi,S)}} q^{\mathrm{dinv}(P)},$$

and call the set of parking squares of (π, S) which have no parking squares of (π, S) above them "top" squares of (π, S), and denote these $\mathrm{top}(\pi, S)$. Note that if $P \in \mathcal{P}_{(\pi,S)}$ contains car $n - |S|$ in a given top square p, we can predict how many inversions will involve car $n - |S|$ without knowing anything else about P (since car $n - |S|$ will be the largest car in P). We call these inversions the "induced inversions" of p, and denote their number by $\mathrm{ind}(\pi, S, p)$. For example, for the path (π, S) in Figure 7, the elements of S are indicated by x's, and there are four top squares, in columns 1, 3, 6 and 8, with ind values 1, 1, 1, and 2, respectively.

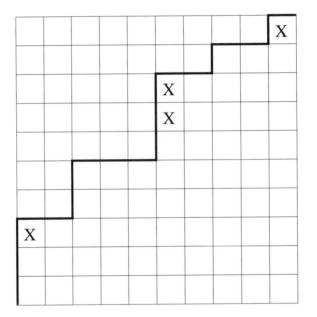

FIGURE 7. A generalized Dyck path.

THEOREM 6.21. *Given* $\pi \in L_{n,n}^+$, *set* $\mathcal{H}_{(\pi,S)}(q) = 1$ *if* $|S| = n$. *Then for* $|S| < n$, $\mathcal{H}_{(\pi,S)}(q)$ *satisfies the recurrence*

$$(6.46) \qquad \mathcal{H}_{(\pi,S)} = \sum_{\text{top squares } p} q^{\mathrm{ind}(\pi,S,p)} \mathcal{H}_{(\pi,S \cup p)},$$

where the sum is over all top squares p *of* (π, S).

PROOF. In any $P \in \mathcal{P}_{(\pi,S)}$, car $n - |S|$ must occur in a top square p, and the number of inversion pairs in P involving this car is by definition $\mathrm{ind}(\pi, S, p)$. Eq. 6.46 is now immediate. □

We now extend the definition of the polynomials $\mathcal{F}_\pi$ to generalized Dyck paths. For $P \in \mathcal{P}_{(\pi,S)}$, define the reading word $\mathrm{read}(P)$ to be the permutation obtained by reading the cars in P along diagonals in a SE direction, outside to in as for ordinary Dyck paths, ignoring all elements of S as they are encountered, and similarly for word parking functions for (π, S). Define symmetric functions $\mathcal{F}_{(\pi,S)}$ via the equations

$$(6.47) \qquad \langle \mathcal{F}_{(\pi,S)}, h_\mu \rangle = \sum_{\substack{P \in \mathcal{P}_{n,(\pi,S)} \\ \mathrm{read}(P) \text{ is a } (\mu, \emptyset)\text{-shuffle}}} q^{\mathrm{dinv}(P)},$$

for all μ. Equivalently we could define the $\mathcal{F}_{(\pi,S)}$ via

$$(6.48) \qquad \mathcal{F}_{(\pi,S)} = \sum_{\sigma \in \mathcal{WP}_{n,(\pi,S)}} x^\sigma q^{\mathrm{dinv}(\sigma)}.$$

Note that $\mathcal{F}_{(\pi,S)}$ is an LLT product of vertical strips, and hence a Schur positive symmetric function.

Given (π, S) and a subset B of $\mathrm{top}(\pi, S)$, assume we have a parking function P for (π, S) with cars $n - |S| - |B| + 1$ through $n - |S|$ occupying the squares of B. If in addition $\mathrm{read}(P)$ is a shuffle of $[n - |S| - |B| + 1, \ldots, n - |S|]$, then we can predict how many inversion pairs in $\mathrm{dinv}(P)$ involve cars in B - call this $\mathrm{ind}(\pi, S, B)$.

THEOREM 6.22. Let (π, S) be a generalized Dyck path. Then $F_{(\pi,\pi)}(X; q) = 1$ and for $|S| < n$,

$$(6.49)$$

$$h_k^\perp F_{(\pi,S)}(X; q) = \begin{cases} \sum_{\substack{B \in \mathrm{top}(\pi,S) \\ |B| = k}} q^{ind(\pi,S,B)} F_{(\pi, S \cup B)}(X; q) & \text{if } 1 \leq k \leq c(\pi, S) \\ 0 & \text{if } c(\pi, S) < k \leq n - |S|. \end{cases}$$

Note that by iterating Theorem 6.22, we get

$$(6.50) \qquad \prod_i h_{\mu_i}^\perp F_{(\pi,S)}(X; q) = \sum_{\substack{P \in \mathcal{P}_{n,(\pi,S)} \\ \mathrm{read}(P) \text{ is a } (\mu, \emptyset)\text{-shuffle}}} q^{\mathrm{dinv}(P)},$$

an equivalent form of (6.47).

LEMMA 6.23. Given a composition $\alpha_1, \ldots, \alpha_m$ of n, let $\pi(\alpha)$ be the balanced Dyck path which, starting at $(0, 0)$, consists of α_1 N steps, followed by α_1 E steps, followed by α_2 N steps, followed by α_2 E steps, ..., followed by α_m N steps followed by α_m E steps. Then

$$(6.51) \qquad mindinv(\pi(\alpha)) = \sum_{1 \leq i < j \leq m} \min(\alpha_i - 1, \alpha_j).$$

PROOF. We refer to column $1 + \alpha_1 + \ldots + \alpha_{i-1}$ of $\pi(\alpha)$ as $\mathrm{Col}_i = \mathrm{Col}_i(\pi(\alpha))$, so all the open squares in $\pi(\alpha)$ are in $\mathrm{Col}_1, \ldots, \mathrm{Col}_m$. Given a parking function for $\pi(\alpha)$, let $\mathrm{occupant}(i, r)$ denote the car in Col_i which is $r - 1$ rows above the diagonal $x = y$. If $\alpha_j \leq \alpha_i - 1$, then for any any $1 \leq r \leq \alpha_j$, either $\mathrm{occupant}(i, r)$ forms an inversion pair with $\mathrm{occupant}(j, r)$, or $\mathrm{occupant}(i, r + 1)$ does. If $\alpha_j \geq \alpha_i$, then for any $2 \leq r \leq \alpha_i$, either occupant(j, r) forms an inversion pair with occupant(i, r), or occupant$(j, r - 1)$ does. In any case we have

$$(6.52) \qquad \text{mindinv}(\pi(\alpha)) \geq \sum_{1 \leq i < j \leq m} \min(\alpha_i - 1, \alpha_j).$$

On the other hand, let γ be the parking function which has cars $1, 2, \ldots, \alpha_1$ in Col$_m$, ..., and cars $\alpha_1 + \ldots + \alpha_{m-1} + 1, \ldots, n$ in Col$_1$. One checks that

$$(6.53) \qquad \text{dinv}(\gamma) = \sum_{1 \leq i < j \leq m} \min(\alpha_i - 1, \alpha_j),$$

so the lower bound on mindinv$(\pi(\alpha))$ is obtained, which establishes (6.51). $\qquad \square$

Proof of Theorem 6.8 In [**GP92**], it is shown that the polynomials

$$(6.54) \qquad \sum_{\beta} K_{\beta, \mu(\alpha)}(q) s_\beta(Z)$$

are characterized by certain skewing operator equations of a form very similar to (6.49). One way of phrasing their result is that if we let μ be a partition of n and we recursively define polynomials $H(z; \pi(\mu))$ associated to the balanced Dyck paths $\pi(\mu)$ by the equations

$$(6.55) \qquad h_k^{\perp} H(Z; \pi(\mu)) = \sum_{|B| = k, B \subseteq \text{top}(\pi(\mu))} q^{\text{stat}(\mu, B)} H(Z; \pi(\mu - B)),$$

then

$$(6.56) \qquad H(Z; \pi(\mu)) = \sum_{\beta} K_{\beta', \mu(\alpha)}(t) s_\beta(Z).$$

In (6.55) $\mu - B$ is the partition obtained by decreasing those parts of μ which correspond to columns with elements of B in them by one, then rearranging into nondecreasing order. The statistic stat$(\mu; B)$ can be expressed as

$$(6.57) \qquad \text{stat}(\mu, B) = \sum_{1 \leq i < j \leq m} g(i, j),$$

where $m = \ell(\mu)$ and

$$(6.58) \qquad g(i, j) = \begin{cases} 1 & \text{if } \mu_i = \mu_j, \ \text{Col}_i(\pi(\mu)) \cap B = \emptyset, \text{ and } \text{Col}_j(\lambda(\mu)) \cap B \neq \emptyset \\ 0 & \text{else.} \end{cases}$$

Next note that if, starting with a composition α, we apply the skewing operator recurrence relations to $F_{\pi(\alpha), X}$, we get the same solution as if we apply them to $F_{\pi(\alpha(X))}$, where $\pi(\alpha(X))$ is the balanced path obtained by starting with the generalized path $(\pi(\alpha), X)$ and removing all rows which contain squares in X, and collapsing in the obvious way, i.e. replacing α by the composition obtained by decreasing each part of α by the number of elements of X in the corresponding column.

In view of this fact and (6.55), (6.56), the theorem will follow by induction if we can show that for all relevant B

$$(6.59) \qquad \text{mindinv}(\pi(\alpha), B) + \text{ind}(\pi(\alpha), B) - \text{mindinv}(\pi(\alpha)) = \text{stat}(\mu(\alpha), B).$$

Let (i, j) satisfy $1 \le i < j \le m$, and let $\text{mindinv}(\pi(\alpha), B; i, j)$, $\text{ind}(\pi(\alpha), B; i, j)$ and $\text{mindinv}(\pi(\alpha); i, j)$ denote the portions of $\text{mindinv}(\pi(\alpha), B)$, $\text{ind}(\pi(\alpha), B)$ and $\text{mindinv}(\pi(\alpha))$ involving Col_i and Col_j of $\pi(\alpha)$, respectively. We will show that

$$(6.60) \quad \text{mindinv}(\pi(\alpha), B; i, j) + \text{ind}(\pi(\alpha), B; i, j) - \text{mindinv}(\pi(\alpha); i, j) = g(i, j).$$

To do this we will break things up into several cases, depending on whether or not any elements of B are in Col_i or Col_j, and depending on how α_i compares to α_j. There are 11 cases which need to be considered separately, all very similar and fairly simple. We will provide details for two of these cases, leaving the rest of the proof as an exercise for the interested reader.

Case 1: $\text{Col}_i \cap B = \emptyset$, $\text{Col}_j \cap B \ne \emptyset$, and $\alpha_i = \alpha_j$.

Since the heights of the two columns are the same, there will be an inversion pair between the bottom square of Col_j and the bottom square of Col_i, which contributes 1 to $\text{ind}(\pi(\alpha), B)$, so $\text{ind}(\pi(\alpha), B; i, j) = 1$. By (6.51) $\text{mindinv}(\pi(\alpha), B; i, j) = \min(\alpha_i - 1, \alpha_j - 1) = \alpha_i - 1$ and $\text{mindinv}(\pi(\alpha); i, j) = \min(\alpha_i - 1, \alpha_j) = \alpha_i - 1$. Since $g(i, j) = 1$, (6.60) becomes $\alpha_i - 1 + 1 - (\alpha_i - 1) = 1$.

Case 2: $\text{Col}_i \cap B \ne \emptyset$, $\text{Col}_j \cap B \ne \emptyset$, and $\alpha_i < \alpha_j$.

We have

$$\text{ind}(\pi(\alpha), B; i, j) = \chi(\alpha_i > 1),$$

$$\text{mindinv}(\pi(\alpha), B; i, j) = \begin{cases} 0 & \text{if } \alpha_i = 1, \\ \min(\alpha_i - 2, \alpha_j - 1) = \alpha_i - 2 & \text{if } \alpha_i > 1, \end{cases}$$

$$\text{mindinv}(\pi(\alpha); i, j) = \min(\alpha_i - 1, \alpha_j) = \alpha_i - 1,$$

$$g(i, j) = 0.$$

Thus (6.60) becomes $0 + 0 - 0 = 0$ if $\alpha_i = 1$ and $\alpha_i - 2 + 1 - (\alpha_i - 1) = 0$ if $\alpha_i > 1$. $\square$

REMARK 6.24. Conjectures 6.1 and 3.8 have the following refinement, which has been tested in Maple for small n, k.

CONJECTURE 6.25. For $1 \le k \le n$,

$$(6.61) \qquad \nabla E_{n,k} = \sum_{\substack{\pi \\ k \text{ rows of area } 0}} t^{area(\pi)} F_\pi(X; q),$$

where the sum is over all Dyck paths π having exactly k rows of length 0.

The m-parameter

Let m be a positive integer. Haiman [**Hai02**] has proved that for any partition λ,

$$(6.62) \qquad \langle \nabla^m e_n, s_\lambda \rangle \in \mathbb{N}[q, t].$$

In fact, he has shown that $\nabla^m e_n$ is the character of a certain S_n-module $DH_n^{(m)}$, where $DH_n^{(1)} = DH_n$.

Define the m-parameter (q, t)-Catalan numbers via

$$(6.63) \qquad \langle \nabla^m e_n, s_{1^n} \rangle.$$

In this section we will describe a conjectured combinatorial formula for $C_n^{(m)}(q,t)$, due to Haiman [**Hai00a**], which extends the dinv form of the combinatorial formula for $C_n(q,t)$. We also describe a bounce version of this conjecture, due to N. Loehr, and a corresponding recurrence relation he found. This section also contains a m-parameter extension of Conjecture 6.1, and more generally, a conjectured combinatorial description of the monomial expansion of $\nabla^m E_{n,k}$.

A m-Dyck path is a lattice path from $(0,0)$ to (mn,n) consisting of N and E steps, never going below the line $x = my$. Let $L_{mn,n}^+$ denote the set of these paths. It is well-known that the cardinality of $L_{mn,n}^+$ equals $\frac{1}{mn+1}\binom{nm}{m}$.

For a given m-path π, and a square w which is above π and contained in the rectangle with vertices $(0,0)$, $(mn,0)$, $(0,n)$ and (mn,n), let $arm(w)$ denote the number of full squares in the same row and strictly right of w, and above π. Similarly, we let $leg(\pi)$ denote the number of squares above π, and in the same column and strictly below w (so w is not counted in its arm or leg).

CONJECTURE 6.26. *(Haiman [**Hai00a**]).*

$$(6.64) \qquad C_n^{(m)}(q,t) = \sum_{\pi \in L_{mn,n}^+} q^{b_m(\pi)} t^{area(\pi)},$$

where $area(\pi)$ is the number of full squares below π and above the line $my = x$, and $b_m(\pi)$ is the number of squares w which are above π, contained in the rectangle with vertices $(0,0)$, $(mn,0)$, $(0,n)$ and (mn,n), and satisfy

$$(6.65) \qquad mleg(w) \le arm(w) \le m(leg(w) + 1).$$

Figure 6.26 shows a typical m-path π, together with the squares which contribute to $b_m(\pi)$ or $area(\pi)$.

REMARK 6.27. Conjecture 6.26 is open for all $m \ge 2$.

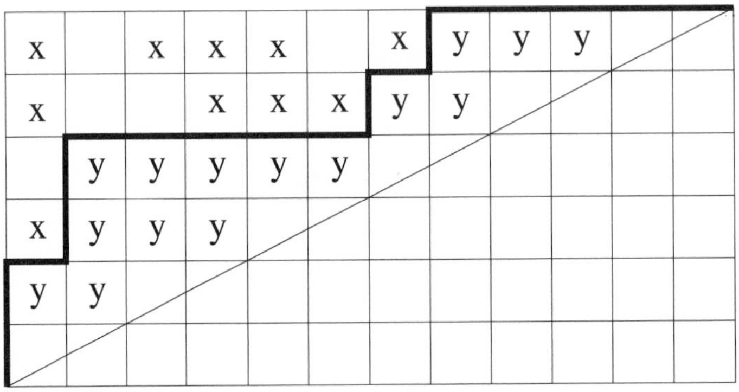

FIGURE 8. A 2-path π with area $= 15$ and $b_m = 10$ (contributors to $b_m(\pi)$ are marked by x's, and to $area(\pi)$ by y's).

We now describe Loehr's extension of the bounce path. To construct the bounce path for an m-path π, start at $(0,0)$ and travel N until you touch an E step of π. Say we traveled N v_1 steps in so doing, then we now travel E v_1 steps. Next we travel N again, say v_2 steps, until we reach an E step of π. Then we travel E $v_1 + v_2$ steps (if $m \ge 2$) or v_1 steps (if $m = 1$). Next we travel N again, say

v_3 steps, until we reach an E step of π. Then we travel E $v_1 + v_2 + v_3$ steps (if $m \geq 3$) or $v_2 + v_3$ steps (if $m = 2$) or v_3 steps (if $m = 1$). In general, after going N and reaching the path, we travel E the sum of the previous m distances we went N, with the understanding that for $k < m$, the distance we travel E on the kth iteration is $v_1 + \ldots + v_k$. To calculate the bounce statistic, each time our bounce path reaches an E step γ of π after travelling N, add up the number of squares above and in the same column as γ. Figure 9 gives the bounce path for a typical 2-path.

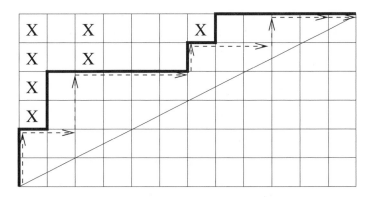

FIGURE 9. A 2-path π with its bounce path (dotted line). The distances $v_1, v_2, \ldots$ the bounce path travels N to reach π are $2, 2, 1, 1, 0$, and so bounce $= 4 + 2 + 1 + 0 + 0 = 7$.

THEOREM 6.28. *(Loehr* [**Loe03**],[**Loe05b**]*)*

$$(6.66) \qquad \sum_{\pi \in L_{mn,n}^+} q^{b_m(\pi)} t^{area(\pi)} = \sum_{\pi \in L_{mn,n}^+} q^{area(\pi)} t^{bounce(\pi)}.$$

EXERCISE 6.29. Consider the algorithm for generating the bounce path for m-Dyck paths.

(1) Show that the bounce path always stays weakly above x=my.
(2) Show that the bounce algorithm always terminates (this means that the bounce path never "gets stuck" in the middle due to a horizontal bounce move being zero).
(3) Show that the area under the bounce path and above $x = my$ is

$$(6.67) \qquad \text{pow} = m \sum_{i \geq 0} \binom{v_i}{2} + \sum_{i \geq 1} v_i \sum_{j=1}^{m} (m - j) v_{i-j}$$

(the v_i's are the vertical moves of the bounce path).

Hint (for all three parts): Compute the (x, y)-coordinates reached by the bounce path after the i'th horizontal move, in terms of the v_i's.

Let $\pi \in L_{mn,n}^+$, and let P be a "parking function for π", i.e. a placement of the integers 1 through n just to the right of N steps of π, with decrease down columns. For a square w containing a car of P, define the coordinates of w to be the (x, y) coordinates of the lower left-hand vertex of w, so for example the bottom square in column 1 has coordinates $(0, 0)$. If $w = (x, y)$, define $\text{diag}(w) = my - x$, so in particular squares on the line $my = x$ have diag value 0. If $w = (x, y)$ contains car i and $w' = (x', y')$ contains car j with $i < j$, we say the pair (w, w') contributes

$$(6.68) \qquad \begin{cases} \max(0, m - |\text{diag}(w') - \text{diag}(w)|) & \text{if } x' > x \\ \max(0, m - |\text{diag}(w') - \text{diag}(w) - 1|) & \text{if } x' < x. \end{cases}$$

Let $\text{dinv}_m(P)$ be the sum of these contributions, over all squares w containing cars. Extend this definition to word parking functions σ by declaring dinv_m to be dinv_m of the standardization of W, where the reading word of W is obtained by reading along diagonals of slope $1/m$, upper right to lower left, outside to in, and standardization is the same as for earlier in the chapter. Let $\mathcal{W}P_n^{(m)}$ denote the set of word parking functions for paths in $L_{mn,n}^+$.

We can now state the m-parameter extension of Conjecture 6.1.

CONJECTURE 6.30 (HHLRU05).

$$(6.69) \qquad \nabla^m e_n = \sum_{\sigma \in \mathcal{W}P_n^{(m)}} x^\sigma q^{dinv_m(\sigma)} t^{area(\sigma)}.$$

REMARK 6.31. There is a way of describing the value of dinv_m of a word parking function without standardization. Say that pairs of squares (w, w') with equal entries contribute a certain number of inversions, namely the minimum of the two quantities in (6.68). Then dinv_m of a word parking function σ is the sum, over all pairs (w, w') of σ, of the number of inversions contributed by (w, w').

REMARK 6.32. If P is the parking function for π whose reading word is $n \cdots 21$, then $\text{dinv}_m(P)$ is an m-parameter dinv-version of $b_m(\pi)$. See [**HHL$^+$05c**, pp. 225-226] for a combinatorial proof that these two versions are equal.

EXERCISE 6.33.   (1) Verify that for $m = 1$ the definition of $\text{dinv}_m(P)$ reduces to our previous definition of dinv, i.e. that for $\pi \in L_{n,n}^+$ and $P \in \mathcal{P}_\pi$, $\text{dinv}_1(P) = \text{dinv}(P)$.
    (2) Given $\pi \in L_{mn,n}^+$ and P a parking function for π, define the "m-magnification" $P^{(m)}$ of P to be the result obtained by replacing each N step of π and the car in it by m copies of this car. See Figure 10 for an example of this when $m = 2$. Prove the following observation of M. Haiman [**Hai00b**]:

$$(6.70) \qquad \text{dinv}_m(P) = \text{dinv}_1(P^{(m)}).$$

Let $\mathcal{W}P_{n,\pi}^{(m)}$ denote the set of word parking functions for a fixed path $\pi \in L_{mn,n}^+$, and set

$$(6.71) \qquad \mathcal{F}_\pi(X; q) = \sum_{\sigma \in \mathcal{W}P_{n,\pi}^{(m)}} x^\sigma q^{dinv_m(\sigma)}.$$

Then we have the following refinement of Conjecture 6.30.

CONJECTURE 6.34 (HHLRU05).

$$(6.72) \qquad \nabla^m E_{n,k} = \sum_{\pi \in L_{mn,n}^+} \mathcal{F}_\pi(X; q) t^{area(\pi)}.$$

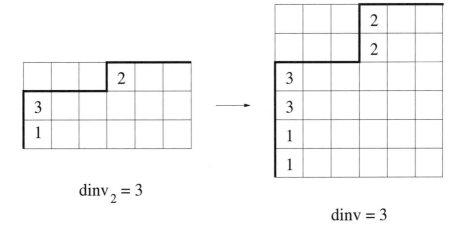

FIGURE 10. The 2-magnification of a parking function.

In [**HHL$^+$05c**] it is shown that $\mathcal{F}_\pi(X;q)$ is a constant power of q, say $e(\pi)$, times an LLT polynomial, specifically an LLT product of vertical columns, where the heights of the columns are the heights of the nonzero columns of π. The columns are permuted with each other, and shifted up and down, in a more complicated way than in the $m=1$ case. As a corollary, by either [**HHL$^+$05c**], the recent work of Grojnowski and Haiman, or the recent work of Assaf, the RHS of (6.69) is a Schur positive symmetric function. The constant $e(\pi)$ can be described as the value of dinv_m obtained by filling all the squares of π with 1's, and using Remark 6.31.

CHAPTER 7

The Proof of the q, t-Schröder Theorem

Summation Formulas for Generalized Skew and Pieri Coefficients

In this section we develop some technical results involving plethystic symmetric function summation formulas, which we will use in the next section to prove Theorem 4.2, by showing that both sides of (4.3) satisfy the same recurrence (see Theorem 4.6) and initial conditions, and are hence equal. This chapter is a condensed version of [**Hag04b**]. In particular, Theorems 7.2, 7.3, 7.6, and 7.9, Corollary 7.3.1, and their proofs, are the same as in [**Hag04b**] with minor variations.

Given $P \in \Lambda$, let $P^*(X) = P[X/M]$, where as usual $M = (1 - q)(1 - t)$. The Cauchy identity (1.67) together with (2.69) implies the useful result

$$(7.1) \qquad s_\lambda^* = \sum_{\beta \vdash |\lambda|} \frac{\tilde{H}_\beta}{w_\beta} \tilde{K}_{\lambda', \beta}(q, t),$$

true for any partition λ. Given $A \in \Lambda$, define generalized Pieri coefficients $d_{\mu\nu}^A$ via the equation

$$(7.2) \qquad A\tilde{H}_\nu = \sum_{\mu \supseteq \nu} \tilde{H}_\mu d_{\mu\nu}^A.$$

For example, (7.1) is equivalent to

$$(7.3) \qquad d_{\beta\emptyset}^{s_\lambda^*} = \frac{\tilde{K}_{\lambda', \beta}}{w_\beta}.$$

Let $A^\perp$ be the skewing operator which is adjoint to multiplication by A with respect to the Hall scalar product. That is, for any symmetric functions A, P, Q,

$$(7.4) \qquad \langle A P, Q \rangle = \langle P, A^\perp Q \rangle.$$

Define skew coefficients $c_{\mu\nu}^{A\perp}$ via

$$(7.5) \qquad A^\perp \tilde{H}_\mu = \sum_{\nu \subseteq \mu} \tilde{H}_\nu c_{\mu\nu}^{A\perp}.$$

The $c_{\mu\nu}^{f\perp}$ and $d_{\mu\nu}^f$ satisfy the important relation [**GH02**, (3.5)].

$$(7.6) \qquad c_{\mu\nu}^{f\perp} w_\nu = d_{\mu\nu}^{\omega f^*} w_\mu.$$

The following result will be crucial in our proof of the q, t-Schröder theorem.

THEOREM 7.1. [**GH02**, pp.698-701] *Let $m \geq d \geq 0$. Then for any symmetric function g of degree at most d, and $\mu \vdash m$,*

$$(7.7) \qquad \sum_{\substack{\nu \subseteq \mu \\ m-d \leq |\nu| \leq m}} c_{\mu\nu}^{(\omega g)^\perp} T_\nu = T_\mu G[D_\mu(1/q, 1/t)],$$

and

$$\sum_{\substack{\nu \subseteq \mu \\ m-d \leq |\nu| \leq m}} c_{\mu\nu}^{g^\perp} = F[D_\mu(q,t)], \tag{7.8}$$

where

$$G[X] = \omega\nabla\left(g\left[\frac{X+1}{(1-1/q)(1-1/t)}\right]\right), \tag{7.9}$$

$$F[X] = \nabla^{-1}\left((\omega g)\left[\frac{X-\epsilon}{M}\right]\right), \tag{7.10}$$

and

$$D_\mu(q,t) = MB_\mu(q,t) - 1. \tag{7.11}$$

We will now use Theorem 7.1 to express $\nabla E_{n,k}$ as a sum over partitions of $n-k$.

THEOREM 7.2. *For $k, n \in \mathbb{N}$ with $1 \leq k < n$,*

$$\nabla E_{n,k} = t^{n-k}(1-q^k)\sum_{\nu \vdash n-k}\frac{T_\nu}{w_\nu}\sum_{\substack{\mu \supseteq \nu \\ \mu \vdash n}}\tilde{H}_\mu\Pi_\mu d_{\mu\nu}^{h_k\left[\frac{X}{1-q}\right]}. \tag{7.12}$$

PROOF. Equations (1.6), (1.11) and (1.20) of [**GH02**] yield

$$\nabla E_{n,k} = \sum_{i=0}^{k}\begin{bmatrix}k\\i\end{bmatrix}(-1)^{n-i}q^{\binom{i}{2}+k-in}\nabla h_n\left[X\frac{1-q^i}{1-q}\right]. \tag{7.13}$$

Plugging the following formula [**GH02**, p. 693]

$$\nabla h_n\left[X\frac{1-q^i}{1-q}\right] = (-t)^{n-i}q^{i(n-1)}(1-q^i)\sum_{\mu \vdash n}\frac{T_\mu\tilde{H}_\mu\Pi_\mu e_i[(1-t)B_\mu(1/q,1/t)]}{w_\mu} \tag{7.14}$$

into (7.13) we get

$$\nabla E_{n,k} = \sum_{i=1}^{k}\begin{bmatrix}k\\i\end{bmatrix}q^{\binom{i}{2}+k-i}(1-q^i)t^{n-i}\sum_{\mu \vdash n}\frac{T_\mu\tilde{H}_\mu\Pi_\mu}{w_\mu}e_i[(1-t)B_\mu(1/q,1/t)]. \tag{7.15}$$

On the other hand, starting with the right-hand side of (7.12),

$$t^{n-k}(1-q^k)\sum_{\nu \vdash n-k}\frac{T_\nu}{w_\nu}\sum_{\substack{\mu \supseteq \nu \\ \mu \vdash n}}\tilde{H}_\mu\Pi_\mu d_{\mu\nu}^{h_k\left[\frac{X}{1-q}\right]} \tag{7.16}$$

$$= t^{n-k}(1-q^k)\sum_{\mu \vdash n}\frac{\tilde{H}_\mu\Pi_\mu}{w_\mu}\sum_{\substack{\nu \subseteq \mu \\ \nu \vdash n-k}}\frac{d_{\mu\nu}^{h_k\left[\frac{X}{1-q}\right]}T_\nu w_\mu}{w_\nu} \tag{7.17}$$

$$= t^{n-k}(1-q^k)\sum_{\mu \vdash n}\frac{\tilde{H}_\mu\Pi_\mu}{w_\mu}\sum_{\substack{\nu \subseteq \mu \\ \nu \vdash n-k}}c_{\mu\nu}^{e_k[(1-t)X]^\perp}T_\nu \tag{7.18}$$

using (7.6). We now apply Theorem 7.1 and use the addition formula (1.66) to obtain

(7.19)

$$
t^{n-k}(1-q^k)\sum_{\mu\vdash n}\frac{\tilde{H}_\mu\Pi_\mu T_\mu}{w_\mu}\omega\nabla\left(h_k\left[\frac{qt(X+1)}{1-q}\right]\right)\Big|_{X=D_\mu(1/q,1/t)}
$$

(7.20)

$$
= t^n q^k(1-q^k)\sum_{\mu\vdash n}\frac{\tilde{H}_\mu\Pi_\mu T_\mu}{w_\mu}\omega\nabla\sum_{j=0}^{k}h_j\left[\frac{X}{1-q}\right]h_{k-j}\left[\frac{1}{1-q}\right]\Big|_{X=D_\mu(1/q,1/t)}
$$

(7.21)

$$
= t^n q^k(1-q^k)\sum_{\mu\vdash n}\frac{\tilde{H}_\mu\Pi_\mu T_\mu}{w_\mu}
$$

$$
\times\sum_{j=0}^{k}q^{\binom{j}{2}}e_j\left[\frac{(1-1/q)(1-1/t)B_\mu(1/q,1/t)-1}{1-q}\right]\frac{1}{(q;q)_{k-j}}
$$

by the $n=\infty$ case of (1.28) and (2.43). After using (1.66) again, (7.21) equals

(7.22)

$$
t^n q^k(1-q^k)\sum_{\mu\vdash n}\frac{\tilde{H}_\mu\Pi_\mu T_\mu}{w_\mu}
$$

$$
\times\sum_{j=0}^{k}q^{\binom{j}{2}-j}t^{-j}\sum_{i=0}^{j}e_i[(1-t)B_\mu(1/q,1/t)]e_{j-i}\left[\frac{-qt}{1-q}\right]\frac{1}{(q;q)_{k-j}}
$$

(7.23)

$$
= t^n q^k(1-q^k)\sum_{\mu\vdash n}\frac{\tilde{H}_\mu\Pi_\mu T_\mu}{w_\mu}\sum_{i=0}^{k}e_i[(1-t)B_\mu(1/q,1/t)]\sum_{j=i}^{k}\frac{q^{\binom{j}{2}-i}t^{-i}(-1)^{j-i}}{(q;q)_{j-i}(q;q)_{k-j}},
$$

since $e_{j-i}[-1/1-q]=(-1)^{j-i}h_{j-i}[1/1-q]=(-1)^{j-i}/(q;q)_{j-i}$. The inner sum in (7.23) equals

(7.24) $$t^{-i}q^{-i}\sum_{s=0}^{k-i}\frac{q^{\binom{s+i}{2}}(-1)^s}{(q;q)_s(q;q)_{k-i-s}}$$

(7.25) $$=\frac{t^{-i}q^{-i}}{(q;q)_{k-i}}\sum_{s=0}^{k-i}\frac{q^{\binom{s}{2}+\binom{i}{2}+is}(-1)^s(1-q^{k-i})\cdots(1-q^{k-i-s+1})}{(q;q)_s}$$

(7.26) $$=\frac{t^{-i}q^{\binom{i}{2}-i}}{(q;q)_{k-i}}\sum_{s=0}^{k-i}\frac{q^{\binom{s}{2}+is+(k-i)s-\binom{s}{2}}(1-q^{-k+i})\cdots(1-q^{-k+i+s-1})}{(q;q)_s}$$

(7.27) $$=\frac{t^{-i}q^{\binom{i}{2}-i}}{(q;q)_{k-i}}\sum_{s=0}^{\infty}\frac{(q^{i-k};q)_s}{(q;q)_s}q^{ks}$$

(7.28) $$=\frac{t^{-i}q^{\binom{i}{2}-i}}{(q;q)_{k-i}}(1-q^i)\cdots(1-q^{k-1}),$$

by (1.23), and plugging this into (7.23) and comparing with (7.15) completes the proof. □

We now derive a summation theorem for the $d_{\mu\nu}^A$ which will be needed to reduce the inner sum in (7.12) to a more useful form. The case $\lambda = 1^{|\nu|}$ is essentially equivalent to Theorem 0.3 of [**GH02**]. The statement involves a new linear operator $\mathcal{K}_P$, where $P \in \Lambda$, which we define on the modified Macdonald basis via

$$(7.29) \qquad \mathcal{K}_P \tilde{H}_\mu = \langle \tilde{H}_\mu, P \rangle \tilde{H}_\mu.$$

Note that

$$(7.30) \qquad \mathcal{K}_{s_\lambda} \tilde{H}_\mu = \tilde{K}_{\lambda,\mu} \tilde{H}_\mu.$$

THEOREM 7.3. *Let $A \in \Lambda^b$, $P \in \Lambda^m$, ν a partition with $|\nu| > 0$, $m \in \mathbb{N}$. Then*

$$(7.31) \qquad \sum_{\substack{\mu \supseteq \nu \\ |\mu|=|\nu|+b}} d_{\mu,\nu}^A P[B_\mu]\Pi_\mu = \Pi_\nu (\Delta_{A[MX]} P\left[\frac{X}{M}\right])[MB_\nu]$$

and

$$(7.32) \qquad \sum_{\substack{\mu \supseteq \nu \\ |\mu|=|\nu|+b}} d_{\mu,\nu}^A P[B_\mu]\Pi_\mu = \Pi_\nu \mathcal{K}_{\omega(P)}\left(\sum_{r=0}^{\min(b,m)} (f)_r e_{m-r}\left[\frac{X}{M}\right]\right)[MB_\nu].$$

Here

$$(7.33) \qquad f = \tau_\epsilon \nabla \tau_1 A$$

with τ the linear operator satisfying $\tau_a A = A[X+a]$, and for any $g \in \Lambda$, $(g)_r$ denotes the portion of g which is $\in \Lambda^r$, i.e. of homogeneous degree r.

PROOF. Our proof is essentially the same as the proof of Theorem 0.3 of [**GH02**], but done in more generality. By definition,

$$(7.34) \qquad \sum_{\mu \supseteq \nu} d_{\mu,\nu}^A \tilde{H}_\mu = A\tilde{H}_\nu.$$

We now evaluate both sides of of (7.34) at $X = 1 + z(MB_\lambda - 1)$ and then apply reciprocity (2.55) to both sides to obtain

$$\sum_{\mu \supseteq \nu} d_{\mu,\nu}^A \frac{\prod_{(i,j)\in\mu}(1 - zq^{a'}t^{l'})\tilde{H}_\lambda[1 + z(MB_\mu - 1)]}{\prod_{(i,j)\in\lambda}(1 - zq^{a'}t^{l'})} =$$

$$(7.35) \qquad A[1 + z(MB_\lambda - 1)]\frac{\prod_{(i,j)\in\nu}(1 - zq^{a'}t^{l'})\tilde{H}_\lambda[1 + z(MB_\nu - 1)]}{\prod_{(i,j)\in\lambda}(1 - zq^{a'}t^{l'})}.$$

Since $|\nu| > 0$ there is a common factor of $1 - z$ (corresponding to $(i,j) = (0,0)$) in the numerator of both sides of (7.35). Canceling this as well as the denominators on both sides and then setting $z = 1$, (7.35) becomes

$$(7.36) \qquad \sum_{\mu \supseteq \nu} d_{\mu,\nu}^A \Pi_\mu \tilde{H}_\lambda[MB_\mu] = A[MB_\lambda]\Pi_\nu \tilde{H}_\lambda[MB_\nu].$$

Since the $\tilde{H}$ form a basis this implies that for any symmetric function G,

$$(7.37) \qquad \sum_{\mu \supseteq \nu} d_{\mu,\nu}^A \Pi_\mu G[MB_\mu] = \Pi_\nu (\Delta_{A[MX]} G)[MB_\nu].$$

Letting $G = P^*$ proves (7.31).

Given $G \in \Lambda^m$, $m \geq 0$, by definition

$$(7.38) \qquad G[X] = \sum_{\beta \vdash m} d^G_{\beta, \emptyset} \tilde{H}_\beta.$$

Thus

$$(7.39) \qquad \sum_{\beta \vdash m} A[MB_\beta] d^G_{\beta, \emptyset} \tilde{H}_\beta = \Delta_{A[MX]} G.$$

Now by Theorem 7.1 we have

$$(7.40) \qquad A[MB_\beta] = \sum_{\alpha \subseteq \beta} c^{f^\perp}_{\beta \alpha},$$

where

$$(7.41) \qquad A = \tau_{-1} \nabla^{-1} \left((\omega f)[\frac{X - \epsilon}{M}] \right),$$

or equivalently

$$(7.42) \qquad \omega f^* = \tau_\epsilon \nabla \tau_1 A.$$

Using (7.6) we get

$$(7.43) \qquad \Delta_{A[MX]} G = \sum_{\beta \vdash m} \tilde{H}_\beta d^G_{\beta \emptyset} \sum_{\alpha \subseteq \beta} c^{f^\perp}_{\beta \alpha}$$

$$(7.44) \qquad = \sum_{r=0}^{m} \sum_{\alpha \vdash m - r} \frac{1}{w_\alpha} \sum_{\substack{\beta \supseteq \alpha \\ \beta \vdash m}} \tilde{H}_\beta d^G_{\beta \emptyset} w_\beta d^{\omega f^*}_{\beta \alpha}.$$

Setting $G = s^*_\lambda$ in (7.44) now yields

$$(7.45) \qquad \Delta_{A[MX]} s^*_\lambda = \sum_{r=0}^{m} \sum_{\alpha \vdash m - r} \frac{1}{w_\alpha} \sum_{\substack{\beta \supseteq \alpha \\ \beta \vdash m}} \tilde{H}_\beta \tilde{K}_{\lambda', \beta} d^{\omega f^*}_{\beta \alpha}$$

$$(7.46) \qquad = \mathcal{K}_{s_{\lambda'}} \left(\sum_{r=0}^{m} \sum_{\alpha \vdash m - r} \frac{1}{w_\alpha} \sum_{\substack{\beta \supseteq \alpha \\ \beta \vdash m}} \tilde{H}_\beta d^{\omega f^*}_{\beta \alpha} \right)$$

$$(7.47) \qquad = \mathcal{K}_{s_{\lambda'}} \left(\sum_{r=0}^{m} \sum_{\alpha \vdash m - r} \frac{1}{w_\alpha} \tilde{H}_\alpha (\omega f^*)_r \right)$$

$$(7.48) \qquad = \mathcal{K}_{s_{\lambda'}} \left(\sum_{r=0}^{\min(b,m)} e^*_{m-r} (\omega f^*)_r \right)$$

using (7.1), and the fact that the degree of ωf^* is at most the degree of A, so $(\omega f^*)_r = 0$ for $r > b$. Plugging (7.48) into (7.39) we see the right-hand sides of (7.31) and (7.32) are equal for $P = s_\lambda$, and the equality for general P follows by linearity. $\qquad \square$

COROLLARY 7.3.1. *Let* $1 \le k < n$ *and* $m \in \mathbb{N}$ *with* $\lambda \vdash m$. *Then*

(7.49)

$$\langle \Delta_{s_\lambda} \nabla E_{n,k}, s_n \rangle$$

$$= t^{n-k} \sum_{\nu \vdash n-k} \frac{T_\nu \Pi_\nu}{w_\nu} \mathcal{K}_{s_{\lambda'}} \left(\sum_{p=1}^{\min(k,m)} \begin{bmatrix} k \\ p \end{bmatrix} q^{\binom{p}{2}} (1-q^p) e_{m-p} \begin{bmatrix} X \\ M \end{bmatrix} h_p \begin{bmatrix} X \\ 1-q \end{bmatrix} \right) [M B_\nu].$$

PROOF. Apply Theorem 7.3 to the inner sum on the right-hand side of (7.12) to get

(7.50) $$\langle \Delta_{s_\lambda} \nabla E_{n,k}, s_n \rangle = t^{n-k}(1-q^k) \sum_{\nu \vdash n-k} \frac{T_\nu \Pi_\nu}{w_\nu} \sum_{\substack{\mu \supseteq \nu \\ \mu \vdash n}} d_{\mu\nu}^{h_k[\frac{X}{1-q}]} \Pi_\mu s_\lambda[B_\mu]$$

(7.51) $$= t^{n-k}(1-q^k) \sum_{\nu \vdash n-k} \frac{T_\nu \Pi_\nu}{w_\nu} \mathcal{K}_{s_{\lambda'}} \left(\sum_{p=0}^{\min(k,m)} (f)_p e_{m-p}^* \right) [M B_\nu],$$

where

(7.52) $$f = \tau_\epsilon \nabla \tau_1 h_k \left[\frac{X}{1-q} \right].$$

By Exercise 7.4, we have

(7.53) $$f = \sum_{p=0}^{k} h_p \left[\frac{X}{1-q} \right] \sum_{j=p}^{k} q^{\binom{j}{2}} (-1)^{j-p} h_{j-p} \left[\frac{1}{1-q} \right] h_{k-j} \left[\frac{1}{1-q} \right],$$

and by Exercise 7.5 the inner sum above equals

(7.54) $$q^{\binom{p}{2}} \begin{bmatrix} k-1 \\ k-p \end{bmatrix},$$

so

(7.55) $$(f)_p = h_p \left[\frac{X}{1-q} \right] q^{\binom{p}{2}} \begin{bmatrix} k-1 \\ k-p \end{bmatrix}.$$

Using this and

(7.56) $$(1-q^k) \begin{bmatrix} k-1 \\ k-p \end{bmatrix} = (1-q^p) \begin{bmatrix} k \\ p \end{bmatrix}$$

in (7.51) completes the proof.   □

EXERCISE 7.4. Show that

(7.57) $$\tau_\epsilon \nabla h_k \left[\frac{X+1}{1-q} \right] = \sum_{p=0}^{k} h_p \left[\frac{X}{1-q} \right] \sum_{j=p}^{k} q^{\binom{j}{2}} (-1)^{j-p} h_{j-p} \left[\frac{1}{1-q} \right] h_{k-j} \left[\frac{1}{1-q} \right].$$

EXERCISE 7.5. Show that

(7.58) $$\sum_{j=p}^{k} q^{\binom{j}{2}} (-1)^{j-p} h_{j-p} \left[\frac{1}{1-q} \right] h_{k-j} \left[\frac{1}{1-q} \right] = q^{\binom{p}{2}} \begin{bmatrix} k-1 \\ k-p \end{bmatrix}.$$

The Proof

In this section we will prove Theorem 4.2. We first note that, by (3.34) and (1.56), for any partition β and $\lambda \vdash n$,

$$(7.59) \qquad \langle \Delta_{s_\beta} \nabla E_{n,n}, s_\lambda \rangle = s_\beta[1, q, \ldots, q^{n-1}] \sum_{T \in SYT(\lambda)} q^{\mathrm{maj}(T)}.$$

THEOREM 7.6. *Let $n, k, d \in \mathbb{N}$ with $1 \leq k \leq n$. Set*

$$(7.60) \qquad F_{n,d,k} = \langle \nabla E_{n,k}, e_{n-d} h_d \rangle.$$

Then

$$(7.61) \qquad F_{n,n,k} = \delta_{n,k},$$

and if $d < n$,

$$(7.62) \qquad F_{n,d,k} = t^{n-k} \sum_{p=\max(1,k-d)}^{\min(k,n-d)} \begin{bmatrix} k \\ p \end{bmatrix} q^{\binom{p}{2}} \sum_{b=0}^{n-k} \begin{bmatrix} p+b-1 \\ b \end{bmatrix} F_{n-k,p-k+d,b}$$

with the initial conditions

$$(7.63) \qquad F_{0,0,k} = \delta_{k,0} \quad and \quad F_{n,d,0} = \delta_{n,0} \delta_{d,0}.$$

PROOF. Eq. (7.61) is equivalent to

$$(7.64) \qquad \langle \nabla E_{n,k}, s_n \rangle = \delta_{n,k}.$$

This follows for $k < n$ by the $\lambda = \emptyset$ case of Corollary 7.3.1 and for $k = n$ by the $\beta = \emptyset$ case of (7.59).

Now assume $0 \leq d < n$. If $k = n$, using (2.44) and (7.59)

$$(7.65) \qquad F_{n,d,k} = \langle \nabla E_{n,n}, e_{n-d} h_d \rangle$$

$$(7.66) \qquad = \langle \Delta_{e_{n-d}} \nabla E_{n,n}, s_n \rangle = e_{n-d}[1, q, \ldots, q^{n-1}]$$

$$(7.67) \qquad = \begin{bmatrix} n \\ n-d \end{bmatrix} q^{\binom{n-d}{2}}$$

by (1.38). Using the initial conditions this agrees with the right-hand side of (7.62).

If $k < n$, by (2.44)

$$(7.68) \qquad F_{n,d,k} = \langle \nabla E_{n,k}, e_{n-d} h_d \rangle = \langle \Delta_{e_{n-d}} \nabla E_{n,k}, s_n \rangle$$

$$(7.69) \qquad = t^{n-k} \sum_{\nu \vdash n-k} \frac{T_\nu \Pi_\nu}{w_\nu} \sum_{p=\max(1,k-d)}^{\min(k,n-d)} \begin{bmatrix} k \\ p \end{bmatrix} q^{\binom{p}{2}} (1-q^p) e_{n-d-p}[B_\nu] h_p[(1-t)B_\nu]$$

by Corollary 7.3.1 with $\lambda = (n-d)$. (In the inner sum in (7.69), $e_{n-d-p}[B_\nu] = 0$ if $n - d - p > n - k$ since B_ν has $n - k$ terms.) Reversing summation (7.69) equals

$$(7.70) \qquad t^{n-k} \sum_{p=\max(1,k-d)}^{\min(k,n-d)} \begin{bmatrix} k \\ p \end{bmatrix} q^{\binom{p}{2}} \sum_{\nu \vdash n-k} \frac{(1-q^p) h_p[(1-t)B_\nu] T_\nu \Pi_\nu e_{n-d-p}[B_\nu]}{w_\nu}$$

$$(7.71) \qquad = t^{n-k} \sum_{p=\max(1,k-d)}^{\min(k,n-d)} \begin{bmatrix} k \\ p \end{bmatrix} q^{\binom{p}{2}} \left\langle \nabla e_{n-k}[X \frac{1-q^p}{1-q}], e_{n-d-p} h_{d+p-k} \right\rangle$$

(by Corollary 3.18 and (2.44))

$$(7.72) \qquad = t^{n-k} \sum_{p=\max(1,k-d)}^{\min(k,n-d)} \begin{bmatrix} k \\ p \end{bmatrix} q^{\binom{p}{2}} \sum_{b=1}^{n-k} \begin{bmatrix} p+b-1 \\ b \end{bmatrix} \langle \nabla E_{n-k,b}, e_{n-d-p} h_{d+p-k} \rangle$$

by letting $z = q^p$ in (3.24). The inner sum in (7.72) equals 0 if $b = 0$, so by the initial conditions we can write (7.72) as

$$(7.73) \qquad t^{n-k} \sum_{p=\max(1,k-d)}^{\min(k,n-d)} \begin{bmatrix} k \\ p \end{bmatrix} q^{\binom{p}{2}} \sum_{b=0}^{n-k} \begin{bmatrix} p+b-1 \\ b \end{bmatrix} F_{n-k,p-k+d,b}.$$

□

Comparing Theorems 7.6 and 4.6, we see that $S_{n,d,k}(q,t)$ and $F_{n,d,k}$ satisfy the same recurence and initial conditions and are hence equal, proving Theorem 4.10. Note that by summing Theorem 4.10 from $k = 1$ to n we obtain Theorem 4.2.

Some Related Results

We now list another technical plethystic summation lemma, and a few of its consequences, taken from [**Hag04b**].

LEMMA 7.7. *Given positive integers m, n, k, a partition $\lambda \vdash m$ and a symmetric function P of homogeneous degree n,*

$$(7.74) \qquad \sum_{\nu \vdash n} \frac{\Pi_\nu \langle \tilde{H}_\nu, P \rangle}{w_\nu} \mathcal{K}_{s_\lambda} \left(\sum_{p=1}^{k} \begin{bmatrix} k \\ p \end{bmatrix} q^{\binom{p}{2}} (1-q^p) e^*_{m-p} h_p [\frac{X}{1-q}] \right) [MB_\nu]$$

$$(7.75) \qquad = \sum_{\mu \vdash m} \frac{(1-q^k) h_k [(1-t) B_\mu] (\omega P) [B_\mu] \Pi_\mu \tilde{K}_{\lambda,\mu}}{w_\mu}.$$

Since κ_{e_m} is the identity operator, if we let $k = 1$ and $\lambda = 1$, then the left-hand side of (7.74) becomes

$$(7.76) \qquad \sum_{\nu \vdash n} \frac{\Pi_\nu M B_\nu e_{m-1}[B_\nu] \langle \tilde{H}_\nu, P \rangle}{w_\nu},$$

and we get the following.

COROLLARY 7.7.1. *For n, m positive integers and $P \in \Lambda^n$,*

$$(7.77) \qquad \langle \Delta_{e_{m-1}} e_n, P \rangle = \langle \Delta_{\omega P} e_m, s_m \rangle.$$

EXAMPLE 7.8. *Letting $m = n+1$ and $P = h_1^n$ in Corollary 7.7.1 we get*

$$(7.78) \qquad \mathcal{H}(DH_n; q, t) = \sum_{\mu \vdash n+1} \frac{M(B_\mu)^{n+1} \Pi_\mu}{w_\mu}.$$

Another consequence of Lemma 7.7 is

THEOREM 7.9. *For $n, k \in \mathbb{N}$ with $1 \le k \le n$*

$$(7.79) \qquad \langle \nabla E_{n,k}, s_n \rangle = \delta_{n,k}.$$

In addition if $m > 0$ and $\lambda \vdash m$

$$(7.80) \qquad \langle \Delta s_\lambda \nabla E_{n,k}, s_n \rangle = t^{n-k} \left\langle \Delta_{h_{n-k}} e_m [X \frac{1-q^k}{1-q}], s_{\lambda'} \right\rangle,$$

or equivalently,

$$(7.81) \qquad \langle \Delta s_\lambda \nabla E_{n,k}, s_n \rangle = t^{n-k} \sum_{\mu \vdash m} \frac{(1-q^k) h_k [(1-t) B_\mu] h_{n-k} [B_\mu] \Pi_\mu \tilde{K}_{\lambda',\mu}}{w_\mu}.$$

PROOF. Eq. (7.79) follows for $k < n$ from the $\lambda = \emptyset$ case of Corollary 7.3.1 and for $k = n$ by the $\beta = \emptyset$ case of (7.59). Next we note that for $m > 0$, (7.80) and (7.81) are equivalent by Corollary 3.18. If $k = n$, (7.80) follows from (7.59), (1.67) and [**Sta99**, p. 363]. For $k < n$, to obtain (7.81) apply Lemma 7.7 with $n = n - k$, $P = s_{1^{n-k}}$ and λ replaced by λ', then use Corollary 7.3.1. $\qquad \square$

Combining Theorem 4.10 and (7.80) we obtain an alternate formula for $S_{n,d,k}(q,t)$.

COROLLARY 7.9.1. *Let n, k be positive integers satisfying $1 \le k \le n$, and let $d \in \mathbb{N}$. Then*

$$(7.82) \qquad S_{n,d,k}(q,t) = t^{n-k} \left\langle \Delta h_{n-k} e_{n-d} [X \frac{1-q^k}{1-q}], s_{n-d} \right\rangle.$$

Note that the $d = 0$ case of Corollary 7.9.1 gives a different formula for $\langle \nabla E_{n,k}, s_{1^n} \rangle$ than given by (3.25).

It is easy to see combinatorially that $S_{n+1,d,1}(q,t) = t^n S_{n,d}(q,t)$. Thus Corollary 7.9.1 also implies

COROLLARY 7.9.2.

$$(7.83) \qquad S_{n,d}(q,t) = \langle \Delta h_n e_{n+1-d}, s_{n+1-d} \rangle.$$

We can now sketch the proof of Proposition 3.9.1 promised in Chapter 3. Let Y be the linear operator defined on the modified Macdonald basis via

$$(7.84) \qquad Y \tilde{H}_\nu = \Pi_\nu \tilde{H}_\nu.$$

Another way to express (7.12) is

$$(7.85) \qquad \nabla E_{n,k} = t^{n-k}(1-q^k) Y \left(\sum_{\nu \vdash n-k} \frac{\tilde{H}_\nu T_\nu h_k [\frac{X}{1-q}]}{w_\nu} \right)$$

$$(7.86) \qquad = t^{n-k}(1-q^k) Y \left(h_{n-k} [\frac{X}{M}] h_k [\frac{X}{1-q}] \right) \quad \text{(using (7.1))},$$

which holds for $1 \le k \le n$. Using (2.43) we get

$$(7.87) \quad \nabla E_{n,n-1} = \frac{t(1-q^{n-1})}{(1-t)(1-q)(q;q)_{n-1}} Y(X \tilde{H}_{n-1})$$

$$(7.88) \qquad = \frac{t(1-q^{n-1})}{(1-t)(1-q)(q;q)_{n-1}} Y(d^{e_1}_{(n),(n-1)} \tilde{H}_n + d^{e_1}_{(n-1,1),(n-1)} \tilde{H}_{n-1,1}).$$

The case $\nu = n - 1$ of [**BGHT99**, Eq. (1.39)] implies

$$(7.89) \qquad d^{e_1}_{(n),(n-1)} = \frac{1-t}{q^{n-1}-t} \quad \text{and} \quad d^{e_1}_{(n-1,1),(n-1)} = \frac{1-q^{n-1}}{t-q^{n-1}}.$$

Plugging these into (7.88) we obtain, after some simplification,

$$(7.90) \qquad \nabla E_{n,n-1} = \frac{t(1-q^{n-1})}{(1-q)} \left(\frac{\tilde{H}_n - \tilde{H}_{(n-1,1)}}{q^{n-1} - t} \right).$$

Proposition 3.9.1 now follows from (2.27), Stembridge's formula for $\tilde{K}_{\lambda,\mu}$ when μ is a hook. □

We note that the positivity of (7.90) is predicted by the "science fiction" conjectures of Bergeron and Garsia [BG99]. Finally, we mention that [Hag04b] also contains a proof of the formula for

$$(7.91) \qquad \langle \nabla e_n, h_{n-d} h_d \rangle$$

predicted by the shuffle conjecture from Chapter 6.

The Combinatorics of Macdonald Polynomials

The Monomial Statistics

Let $\mu \vdash n$. We let $\mathrm{dg}(\mu)$ denote the "augmented" diagram of μ, consisting of μ together with a row of squares below μ, referred to as the *basement*, with coordinates $(j, 0)$, $1 \leq j \leq \mu_1$. Define a *filling* σ of μ to be an assignment of a positive integer to each square of μ. For $s \in \mu$, we let $\sigma(s)$ denote the integer assigned to s, i.e the integer occupying s. Define the *reading word* $\sigma_1 \sigma_2 \cdots \sigma_n$ to be the occupants of μ read across rows left to right, starting with the top row and working downwards.

For each filling σ of μ we associate x, q and t weights. The x weight is defined in a similar fashion to SSYT, namely

$$(\mathrm{A.1}) \qquad x^\sigma = \prod_{s \in \mu} x_{\sigma(s)}.$$

For $s \in \mu$, let North(s) denote the square right above s in the same column (we view μ as embedded in the xy-plane, so the square above s always exists, although it may not be in μ), and South(s) the square of $\mathrm{dg}(\mu)$ directly below s, in the same column. Let the decent set of σ, denoted $\mathrm{Des}(\sigma, \mu)$, be the set of squares $s \in \mu$ where $\sigma(s) > \sigma(\mathrm{South}(s))$. (We regard the basement as containing virtual infinity symbols, so no square in the bottom row of σ can be in $\mathrm{Des}(\sigma, \mu)$.) Finally set

$$(\mathrm{A.2}) \qquad \mathrm{maj}(\sigma, \mu) = \sum_{s \in \mathrm{Des}(\sigma, \mu)} \mathrm{leg}(s) + 1.$$

Note that $\mathrm{maj}(\sigma, 1^n) = \mathrm{maj}(\sigma)$, where σ is viewed as a word, and maj is as in (1.8).

We say a square $u \in \mu$ attacks all other squares $v \in \mu$ in its row and strictly to its right, and all other squares $v \in \mu$ in the row below and strictly to its left. We say u, v attack each other if u attacks v or v attacks u. An *inversion pair* of σ is a pair of squares u, v where u attacks v and $\sigma(u) > \sigma(v)$. Let $\mathrm{Invset}(\sigma, \mu)$ denote the set of inversion pairs of σ, $\mathrm{Inv}(\sigma, \mu) = |\mathrm{Invset}(\sigma, \mu)|$ its cardinality and set

$$(\mathrm{A.3}) \qquad \mathrm{inv}(\sigma, \mu) = \mathrm{Inv}(\sigma, \mu) - \sum_{s \in \mathrm{Des}(\sigma, \mu)} \mathrm{arm}(s).$$

For example, if σ is the filling on the left in Figure 1 then

$$\mathrm{Des}(\sigma) = \{(1, 2), (1, 4), (2, 3), (3, 2)\},$$
$$\mathrm{maj}(\sigma) = 3 + 1 + 2 + 1 = 7,$$
$$\mathrm{Invset}(\sigma) = \{[(1, 4), (2, 4)], [(2, 4), (1, 3)], [(2, 3), (1, 2)], [(1, 2), (3, 2)],$$
$$[(2, 2), (3, 2)], [(2, 2), (1, 1)], [(3, 2), (1, 1)], [(2, 1), (3, 1)], [(2, 1), (4, 1)]\}$$
$$\mathrm{inv}(\sigma) = 9 - (2 + 1 + 0 + 0) = 6.$$

Note that $\mathrm{inv}(\sigma, (n)) = \mathrm{inv}(\sigma)$, where σ is viewed as a word and inv is as in (1.8).

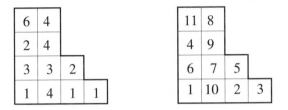

FIGURE 1. On the left, a filling of $(4, 3, 2, 2)$ with reading word 64243321411 and on the right, its standardization.

DEFINITION A.1. For $\mu \vdash n$, let

$$(A.4) \qquad C_\mu(X; q, t) = \sum_{\sigma:\mu\to\mathbb{Z}^+} x^\sigma t^{\mathrm{maj}(\sigma,\mu)} q^{\mathrm{inv}(\sigma,\mu)}.$$

We define the standardization of a filling σ, denoted σ', as the standard filling whose reading word is the standardization of $\mathrm{read}(\sigma)$. Figure 1 gives an example of this. It is immediate from Definition A.1 and Remark 6.13 that

$$(A.5) \qquad C_\mu(X; q, t) = \sum_{\beta\in S_n} t^{\mathrm{maj}(\beta,\mu)} q^{\mathrm{inv}(\beta,\mu)} Q_{n,\mathrm{Des}(\beta^{-1})}(X),$$

where we identity a permutation β with the standard filling whose reading word is β.

REMARK A.2. There is another way to view $\mathrm{inv}(\sigma, \mu)$ which will prove useful to us. Call three squares u, v, w, with $u, v \in \mu$, $w = \mathrm{South}(u)$, and with v in the same row as u and strictly to the right of μ, a *triple*. Given a standard filling σ, we define an orientation on such a triple by starting at the square, either u, v or w, with the smallest element of σ in it, and going in a circular motion, towards the next largest element, and ending at the largest element. We say the triple is an inversion triple or a coinversion triple depending on whether this circular motion is counterclockwise or clockwise, respectively. Note that since $\sigma(j, 0) = \infty$, if u, v are in the bottom row of σ, they are part of a counterclockwise triple if and only if $\sigma(u) > \sigma(v)$. Extend this definition to (possibly non-standard) fillings by defining the orientation of a triple to be the orientation of the corresponding triple for the standardized filling σ'. (So for two equal numbers, the one which occurs first in the reading word is regarded as being smaller.) It is an easy exercise to show that $\mathrm{inv}(\sigma, \mu)$ is the number of counterclockwise triples. For example, for the filling in Figure 1, the inversion triples are

$$(A.6) \qquad [(1,3),(1,4),(2,4)], [(1,2),(1,3),(2,3)], [(1,1),(1,2),(3,2)],$$
$$[(2,1),(2,2),(3,2)], [(2,1),(3,1)], [(2,1),(4,1)].$$

The following theorem was conjectured by Haglund [**Hag04a**] and proved by Haglund, Haiman and Loehr [**HHL05b**], [**HHL05a**]. It gives a combinatorial formula for the $\tilde{H}_\mu$.

THEOREM A.3. *For $\mu \in Par$,*

$$(A.7) \qquad C_\mu[X; q, t] = \tilde{H}_\mu[X; q, t].$$

EXAMPLE A.4. Theorem's A.3 and 2.8 together imply (see also remark 1.30)

$$(A.8) \qquad \mathcal{H}(V(\mu); q, t) = \sum_{\sigma \in S_n} t^{\mathrm{maj}(\sigma, \mu)} q^{\mathrm{inv}(\sigma, \mu)}.$$

EXERCISE A.5. Let $F_0(\mu) = 1$ for all μ, $F_k(\emptyset) = 1$ for all k, and for $k > 0$ and $|\mu| > 0$ define

$$F_k(\mu) = \sum_{\nu \to \mu} c_{\mu, \nu} F_{k-1}(\nu),$$

where $c_{\mu, \nu} = c_{\mu, \nu}^{e_1 \perp}$ as in (7.5), and $\nu \to \mu$ means $\nu \subset \mu$ and $|\nu| = |\mu| - 1$. For $k = 1$ the right-hand side of this identity equals B_μ [**BGHT99**]. Garsia and Haiman [**GH98**, p. 107] showed that for $\mu \vdash n$, $F_{n-1}(\mu) = \langle \tilde{H}_\mu, h_{1^n} \rangle$. Show more generally that

$$F_k(\mu) = \langle \tilde{H}_\mu, h_{n-k} h_{1^k} \rangle.$$

OPEN PROBLEM A.6. *Theorem 2.8 (or (2.30) implies the well-known symmetry relation*

$$(A.9) \qquad \tilde{H}_\mu[Z; q, t] = \tilde{H}_{\mu'}[Z; t, q].$$

(This can be derived fairly easily from the three axioms in Theorem A.8 below.) Prove this symmetry combinatorially using Theorem A.3. Note that in the case $\mu = (n)$ this question is equivalent to asking for a bijective proof that maj and inv have the same distribution on arbitrary multisets, which is exactly what Foata's ϕ map from Chapter 1 gives.

OPEN PROBLEM A.7. *Garsia and Haiman have conjectured the existence of symmetric functions $\tilde{H}_L(X; q, t)$ which can be associated to any set L of squares in the first quardant of the xy-plane, and which satisfy certain defining axioms [**GH95**]; see also [**BBG$^+$99**]. If L is a partition diagram μ, then $\tilde{H}_L(X; q, t)$ equals the Macdonald polynomial $\tilde{H}_\mu(X; q, t)$. In a recent Ph. D. thesis under Garsia, Jason Bandlow [**Ban07**] has shown that for skew shapes with no more than two squares in any column, the obvious extension of formula (A.4) correctly generates $\tilde{H}_L(X; q, t)$. His proof reveals some refined properties of the inv and maj statistics which hold for two-row partition shapes. For more general sets of squares though, Bandlow has found that the obvious extension of (A.4) doesn't generate the $\tilde{H}_L$. Find a way to modify the inv and maj statistics to correctly generate the $\tilde{H}_L$ for general L.*

Proof of the Formula

Theorem 2.2, when translated into a statement about the $\tilde{H}_\mu$ using (2.45) gives [**Hai99**], [**HHL05a**]

THEOREM A.8. *The following three conditions uniquely determine a family $\tilde{H}_\mu(X; q, t)$ of symmetric functions.*

$$(A.10) \qquad \tilde{H}_\mu[X(q - 1); q, t] = \sum_{\rho \leq \mu'} c_{\rho, \mu}(q, t) m_\rho(X)$$

$$(A.11) \qquad \tilde{H}_\mu[X(t - 1); q, t] = \sum_{\rho \leq \mu} d_{\rho, \mu}(q, t) m_\rho(X)$$

$$(A.12) \qquad \tilde{H}_\mu(X; q, t)|_{x_1^n} = 1.$$

In this section we show that $C_\mu(X; q, t)$ satisfies the three axioms above, and is hence equal to $\tilde{H}_\mu$. It is easy to see that $C_\mu(X; q, t)$ satisfies (A.12), since the only way to get an x-weight of x_1^n is a filling of all 1's, which has no inversion triples and no descents.

Next we argue that C_μ can be written as a sum of LLT polynomials as defined in Chapter 6. Fix a descent set D, and let

$$(A.13) \qquad F_D(X; q) = \sum_{\substack{\sigma: \mu \to \mathbb{Z}^+ \\ \mathrm{Des}(\sigma, \mu) = D}} q^{\mathrm{inv}(\beta, \mu)} x^\sigma.$$

If μ has one column, then F_D is a ribbon Schur function. More generally, $F_D(X; q)$ is an LLT product of ribbons. We illustrate how to transform a filling σ into a term $\gamma(\sigma)$ in the corresponding LLT product in Figure 2. The shape of the ribbon corresponding to a given column depends on the descents in that column; if $s \in \mathrm{Des}(\sigma, \mu)$, then there is a square in the ribbon directly below the square γ_s corresponding to s in $\gamma(\sigma)$, otherwise, there is a square in the ribbon directly to the right of γ_s. Note that inversion pairs in σ are in direct correspondence with LLT inversion pairs in $\gamma(\sigma)$. Thus

$$(A.14) \qquad C_\mu(X; q, t) = \sum_D t^L q^{-A} F_D(X; q),$$

where the sum is over all possible descent sets D of fillings of μ, with

$$L = \sum_{s \in D} \mathrm{leg}(s) + 1$$

$$A = \sum_{s \in D} \mathrm{arm}(s).$$

Since LLT polynomials are symmetric functions, we have

THEOREM A.9. *For all μ, $C_\mu(X; q, t)$ is a symmetric function.*

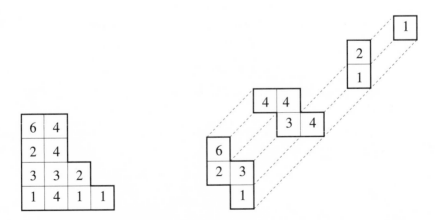

FIGURE 2. On the left, a filling, and on the right, the term in the corresponding LLT product of ribbons.

Since C_μ is a symmetric function, by (A.5), (6.25) and the comments above it, we have

$$(A.15) \qquad \omega^W C_\mu[Z + W; q, t] = \sum_{\beta \in S_n} t^{\mathrm{maj}(\beta,\mu)} q^{\mathrm{inv}(\beta,\mu)} \tilde{Q}_{n,\mathrm{Des}(\beta^{-1})}(Z,W).$$

Also, any symmetric function $f(X)$ of homogeneous degree n is uniquely determined by its value on $X_n = \{x_1, \ldots, x_n\}$ so in Definition A.1 we may as well restrict ourselves to fillings from the alphabet $\mathcal{A}_+ = \{1, 2, \ldots, n\}$, and in (A.15) to fillings from the alphabet $\mathcal{A}_\pm = \{1, 2, \ldots, n, \bar{1}, \ldots, \bar{n}\}$. In Chapter 6 we were assuming that the elements of $\mathcal{A}_\pm$ satisfied the total order relation

$$(A.16) \qquad\qquad 1 < \bar{1} < 2 < \bar{2} < \cdots < n < \bar{n},$$

but (A.15) actually holds no matter which total ordering of the elements we use. Our proof that $\tilde{C}_\mu$ satisfies axiom (A.10) assumes the ordering (A.16) while our proof that it satisfies axiom (A.11) uses the ordering

$$(A.17) \qquad\qquad 1 < 2 < \cdots < n < \bar{n} < \cdots < \bar{2} < \bar{1}.$$

Now for a real parameter α, $C_\mu[X(\alpha-1); q, t]$ can be obtained from $\omega^W C_\mu[Z + W; q, t]$ by replacing z_i by αx_i and w_i by $-x_i$, for $1 \le i \le n$ (since $\omega f(X) = (-1)^n f[-X]$). Hence (A.15) implies

$$(A.18) \qquad C_\mu[X\alpha - X; q, t] = \sum_{\tilde{\sigma}: \mu \to \mathcal{A}_\pm} x^{|\tilde{\sigma}|} t^{\mathrm{maj}(\tilde{\sigma},\mu)} q^{\mathrm{inv}(\tilde{\sigma},\mu)} \alpha^{\mathrm{pos}(\tilde{\sigma})} (-1)^{\mathrm{neg}(\tilde{\sigma})}$$

where $|\tilde{\sigma}|$ is the filling obtained by replacing each negative letter $\bar{j}$ by j for all $1 \le j \le n$, and leaving the positive letters alone. Here pos and neg denote the number of positive letters and negative letters in σ, respectively. We set

$$(A.19) \qquad\qquad \mathrm{maj}(\tilde{\sigma}) = \mathrm{maj}((\tilde{\sigma})')$$

$$(A.20) \qquad\qquad \mathrm{inv}(\tilde{\sigma}) = \mathrm{inv}((\tilde{\sigma})').$$

Here the standardization $(\tilde{\sigma})'$ is constructed as in Chapter 6 (below Open Problem 6.11), where for two equal positive letters, the one that occurs first in the reading word is regarded as smaller, while for two equal negative letters, the one that occurs first is regarded as larger. The standardization process also depends on the total order on $\mathcal{A}_\pm$ we choose though, as in Figure 3.

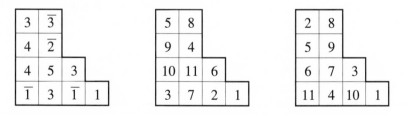

FIGURE 3. On the left, a super filling. In the middle, its standardization assuming the ordering (A.16) and on the right, its standardization assuming the ordering (A.17).

Define the *critical pair* of squares of $\tilde{\sigma}$ as the last pair (in the reading word order) of attacking squares u, v which both contain elements of the set $\{1, \bar{1}\}$, with u say denoting the "critical square", i.e. the earliest of these two squares in the reading order. If $\tilde{\sigma}$ has no such attacking pair, then we define the critical pair as the last pair of attacking squares which both contain elements of the set $\{2, \bar{2}\}$, and if no such pair exists, as the last pair of attacking squares which both contain elements of the set $\{3, \bar{3}\}$, etc. Let $IM(\tilde{\sigma})$ be the sign-reversing involution obtained by starting with $\tilde{\sigma}$ and switching the sign of the element in the critical square. For example, for the filling on the left in Figure 3, the critical pair is $(3, 1), (4, 1)$ and M changes the $\bar{1}$ in $(3, 1)$ to a 1.

Assume our alphabets satisfy (A.16), and let $\alpha = q$ in (A.18). Clearly $IM(\tilde{\sigma})$ fixes the x weight. Let u, v be critical with $|\tilde{\sigma}(u)| = a$. If $|\tilde{\sigma}(\text{North}(u))| = a$ we have $u \in \text{Des}(\tilde{\sigma}, \mu)$ if and only if $\tilde{\sigma}(\text{North}(u)) = \bar{a}$. Furthermore if u is not in the bottom row of μ, then $|\tilde{\sigma}(\text{South}(u))| \neq a$, since otherwise v and $\text{South}(u)$ are an attacking pair later in the reading order, and u, v would not be the critical pair. It follows that $IM(\tilde{\sigma})$ fixes the t weight also.

Since $IM(\tilde{\sigma})$ fixes the descent set, it will fix the q-weight if and only if changing a to $\bar{a}$ increases the number of inversion pairs Inv by 1 (since this change decreases pos by 1). Changing a to $\bar{a}$ has no affect on inversion pairs involving numbers not equal to a or $\bar{a}$. The definition of critical implies that changing a to $\bar{a}$ creates a new inversion pair between u and v, but does not affect any other inversion pairs. Hence if $\alpha = q$ in (A.18), then $M(\tilde{\sigma})$ fixes the q-weight too.

Call a super filling nonattacking if there are no critical pairs of squares. (These are the fixed points of IM). The above reasoning shows

(A.21)
$$C_\mu[X(q-1); q, t] = \sum_{\substack{\tilde{\sigma}: \mu \to \mathcal{A}_\pm, \text{ nonattacking}}} x^{|\tilde{\sigma}|} t^{\text{maj}(\tilde{\sigma}, \mu)} q^{\text{inv}(\tilde{\sigma}, \mu)} q^{\text{pos}(\tilde{\sigma})} (-1)^{\text{neg}(\tilde{\sigma})},$$

where the sum is over all nonattacking superfillings of μ. Since nonattacking implies there is at most one occurrence of a number from the set $\{a, \bar{a}\}$ in any row, the exponent of x_1 in $x^{|\tilde{\sigma}|}$ is at most μ_1', the sum of the exponents of x_1 and x_2 is at most $\mu_1' + \mu_2'$, etc., which shows C_μ satisfies axiom (A.10).

Call the first square w (in the reading word order) such that $|\tilde{\sigma}(w)| = 1$, with w not in the bottom row, the *pivotal square*. If there is no such square, let w denote the first square, not in the bottom *two* rows of μ, with $|\tilde{\sigma}(w)| = 2$, and again if there is no such square, search for the first square not in the bottom three rows satisfying $|\tilde{\sigma}(w)| = 3$, etc. Let $IN(\tilde{\sigma})$ denote the sign-reversing involution obtained by switching the sign on $\tilde{\sigma}(w)$. For the filling on the left in Figure 3, the pivotal square is $(2, 3)$, and IN switches this $\bar{2}$ to a 2.

Let our alphabets satisfy (A.17), with $\alpha = t$ in (A.18). As before, $IN(\tilde{\sigma})$ trivially fixes the x-weight. Note that by construction, if the pivotal square is in row k, then after standardization it contains either the smallest element occurring anywhere in rows $k-1$ and higher or the largest element occurring in rows $k-1$ and higher. Thus in any triple containing the pivotal square, IN either switches an element smaller than the other to an element larger than the other two, or vice-versa, which preserves the orientation of the triple. By Remark A.2, IN thus fixes the q-weight.

If IN switches an element in row k, it follows that it switches the globally smallest element in rows $k-1$ and higher to the globally largest element in rows $k-1$ and higher. Assume for the moment that $\text{North}(w) \in \mu$. Then $\text{North}(w)$ will be in the descent set if and only if $\tilde{\sigma}(w)$ is positive. Similarly, since by construction

South$(w) \in \mu$, w will be in Des$(\tilde{\sigma}, \mu)$ if and only if $\tilde{\sigma}(w)$ is negative. It follows that switching $\tilde{\sigma}(w)$ from a positive to the corresponding negative letter increases the value of maj by 1, which also holds by inspection if North$(w) \notin \mu$. Now let $\alpha = t$ in (A.18). Since the positive choice for $\tilde{\sigma}(w)$ has an extra t power coming from the t^{pos} term, it follows that $IN(\tilde{\sigma})$ fixes the t weight.

Call a super filling *primary* if there are no pivotal squares. (These are the fixed points of IN). We have

$$(A.22) \quad C_\mu[X(t-1); q, t] = \sum_{\tilde{\sigma}: \mu \to \mathcal{A}_\pm, \text{primary}} x^{|\tilde{\sigma}|} t^{\text{maj}(\tilde{\sigma}, \mu)} q^{\text{inv}(\tilde{\sigma}, \mu)} t^{\text{pos}(\tilde{\sigma})} (-1)^{\text{neg}(\tilde{\sigma})},$$

where the sum is over all primary superfillings of μ. Now primary implies that all 1 or $\bar{1}$'s are in the bottom row, all 2 or $\bar{2}$'s are in the bottom two rows, etc., so in $x^{|\tilde{\sigma}|}$ the power of x_1 is at most μ_1, the sum of the powers of x_1 and x_2 is at most $\mu_1 + \mu_2$, etc. Thus C_μ satisfies axiom A.11, which completes the proof of Theorem A.3. $\square$.

EXERCISE A.10. Use (A.18) to prove the $\lambda = \emptyset$ case of (2.55).

Consequences of the Formula

The Cocharge Formula for Hall-Littlewood Polynomials. In this subsection we show how to derive (1.53), Lascoux and Schützenberger's formula for the Schur coefficients of the Hall-Littlewood polynomials, from Theorem A.3. This application was first published in [**HHL05b**] and [**HHL05a**], although the exposition here is taken mainly from [**Hag06**].

We require the following lemma, whose proof is due to N. Loehr and G. Warrington [**LW03**].

LEMMA A.11. *Let $\mu \vdash n$. Given multisets M_i, $1 \leq i \leq \ell(\mu)$, of positive integers with $|M_i| = \mu_i$, there is a unique filling σ with the property that $\text{inv}(\sigma, \mu) = 0$, and for each i in the range $1 \leq i \leq \ell(\mu)$, the multiset of elements of σ in the ith row of μ is M_i.*

PROOF. Clearly the elements in the bottom row will generate no inversions if and only if they are in monotone nondecreasing order in the reading word. Consider the number to place in square $(1, 2)$, i.e. right above the square $(1, 1)$. Let p be the smallest element of M_2 which is strictly bigger than $\sigma(1, 1)$, if it exists, and the smallest element of M_2 otherwise. Then if $\sigma(1, 2) = p$, one sees that $(1, 1)$ and $(1, 2)$ will not form any inversion triples with $(j, 2)$ for any $j > 1$. We can iterate this procedure. In square $(2, 2)$ we place the smallest element of $M_2 - \{p\}$ (the multiset obtained by removing one copy of p from M_2) which is strictly larger than $\sigma(2, 1)$, if it exists, and the smallest element of $M_2 - \{p\}$ otherwise, and so on, until we fill out row 2. Then we let $\sigma(1, 3)$ be the smallest element of M_3 which is strictly bigger than $\sigma(1, 2)$, if it exists, and the smallest element of M_3 otherwise, etc. Each square (i, j) cannot be involved in any inversion triples with $(i, j-1)$ and (k, j) for some $k > i$, so $\text{inv}(\sigma) = 0$. For example, if $M_1 = \{1, 1, 3, 6, 7\}$, $M_2 = \{1, 2, 4, 4, 5\}$, $M_3 = \{1, 2, 3\}$ and $M_4 = \{2\}$, then the corresponding filling with no inversion triples is given in Figure 4. $\square$

Given a filling σ, we construct a word cword(σ) by initializing cword to the empty string, then scanning through read(σ), from the beginning to the end, and each time we encounter a 1, adjoin the number of the row containing this 1 to the beginning of cword. After recording the row numbers of all the 1's in this fashion, we go back to the beginning of read(σ), and adjoin the row numbers of squares containing 2's to the beginning of cword. For example, if σ is the filling in Figure 4, then cword(σ) = 11222132341123.

2				
3	1	2		
2	4	4	1	5
1	1	3	6	7

FIGURE 4. A filling with no inversion triples.

Assume σ is a filling with no inversion triples, We translate the statistic maj(σ, μ) into a statistic on cword(σ). Note that $\sigma(1,1)$ corresponds to the rightmost 1 in cword(σ) - denote this 1 by w_{11}. If $\sigma(1,2) > \sigma(1,1)$, $\sigma(1,2)$ corresponds to the rightmost 2 which is left of w_{11}, otherwise it corresponds to the rightmost 2 (in cword(σ)). In any case denote this 2 by w_{12}. More generally, the element in cword(σ) corresponding to $\sigma(1,i)$ is the first i encountered when travelling left from $w_{1,i-1}$, looping around and starting at the beginning of cword(σ) if necessary. To find the subword $w_{21}w_{22}\cdots w_{2\mu'_2}$ corresponding to the second column of σ, we do the same algorithm on the word cword(σ)' obtained by removing the elements $w_{11}w_{12}\cdots w_{1\mu'_1}$ from cword(σ), then remove $w_{21}w_{22}\cdots w_{2\mu'_2}$ and apply the same process to find $w_{31}w_{32}\cdots w_{3\mu'_3}$ etc..

Clearly $\sigma(i,j) \in \mathrm{Des}(\sigma, \mu)$ if and only if w_{ij} occurs to the left of $w_{i,j-1}$ in cword(σ). Thus maj(σ, μ) is transparently equal to the statistic cocharge(cword(σ)) described in Algorithm 1.22.

We associate a two-line array $A(\sigma)$ to a filling σ with no inversions by letting the upper row $A_1(\sigma)$ be nonincreasing with the same weight as σ, and the lower row $A_2(\sigma)$ be cword(σ). For example, to the filling in Figure 4 we associate the two-line array

$$7\,6\,5\,4\,4\,3\,3\,2\,2\,2\,1\,1\,1\,1$$

(A.23) $$1\,1\,2\,2\,2\,1\,3\,2\,3\,4\,1\,1\,2\,3$$

By construction, below equal entries in the upper row the entries in the lower row are nondecreasing. Since C_μ is a symmetric function, we can reverse the variables, replacing x_i by x_{n-i+1} for $1 \le i \le n$, without changing the sum. This has the effect of reversing $A_1(\sigma)$, and we end up with an ordered two-line array as in Theorem 1.23. We can invert this correspondence since from the two-line array we get the multiset of elements in each row of σ, which uniquely determines σ by Lemma A.11. Thus

(A.24) $$C_\mu(x_1, x_2, \ldots, x_n; 0, t) = \sum_\sigma x^{\mathrm{weight}(A_1(\sigma))} t^{\mathrm{cocharge}(A_2(\sigma))}$$

(A.25) $$= \sum_{(A_1, A_2)} x^{\mathrm{weight}(A_1)} t^{\mathrm{cocharge}(A_2)},$$

where the sum is over ordered two-line arrays satisfying weight(A_2) = μ.

Now it is well known that for any word w of partition weight, cocharge $(w) =$ cocharge $(\text{read}(P_w))$, where $\text{read}(P_w)$ is the reading word of the insertion tableau P_w under the RSK algorithm [**Man01**, pp.48-49], [**Sta99**, p.417]. Hence applying Theorem 1.23 to (A.25),

$$\text{(A.26)} \qquad C_\mu(x_1, x_2, \ldots, x_n; 0, t) = \sum_{(P,Q)} x^{\text{weight}(Q)} t^{\text{cocharge}(\text{read}(P))},$$

where the sum is over all pairs (P, Q) of SSYT of the same shape with $\text{weight}(P) = \mu$. Since the number of different Q tableau of weight ν matched to a given P tableau of shape λ is the Kostka number $K_{\lambda,\nu}$,

$$C_\mu[X; 0, t] = \sum_\nu m_\nu \sum_\lambda \sum_{\substack{P \in SSYT(\lambda,\mu) \\ Q \in SSYT(\lambda,\nu)}} t^{\text{cocharge}(\text{read}(P))}$$

$$= \sum_\lambda \sum_{P \in SSYT(\lambda,\mu)} t^{\text{cocharge}(\text{read}(P))} \sum_\nu m_\nu K_{\lambda,\nu}$$

$$\text{(A.27)} \qquad = \sum_\lambda s_\lambda \sum_{P \in SSYT(\lambda,\mu)} t^{\text{cocharge}(\text{read}(P))}. \qquad \square$$

Formulas for J_μ. By (2.45) we have

$$\text{(A.28)} \qquad J_\mu(Z; q, t) = t^{n(\mu)} \tilde{H}_\mu[Z(1-t); q, 1/t]$$

$$= t^{n(\mu)} \tilde{H}_\mu[Zt(1/t - 1); q, 1/t]$$

$$\text{(A.29)} \qquad = t^{n(\mu)+n} \tilde{H}_{\mu'}[Z(1/t - 1); 1/t, q]$$

using (A.9). Formula (A.21), with q, t interchanged thus implies

$$\text{(A.30)} \quad J_\mu(Z; q, t) = \sum_{\text{nonattacking super fillings } \tilde{\sigma} \text{ of } \mu'} z^{|\tilde{\sigma}|} q^{\text{maj}(\tilde{\sigma},\mu')} t^{\text{coinv}(\tilde{\sigma},\mu')} (-t)^{\text{neg}(\tilde{\sigma})}$$

where $\text{coinv} = n(\mu) - \text{inv}$ is the number of coinversion triples, and we use the ordering (A.16).

We can derive a more compact form of (A.30) by grouping together all the 2^n super fillings $\tilde{\sigma}$ whose absolute value equals a fixed positive filling σ. A moments thought shows that if $|\tilde{\sigma}(s)| > |\tilde{\sigma}(\text{South}(s))|$, then $\tilde{\sigma}(s) > \tilde{\sigma}(\text{South}(s))$. In other words, $\text{Des}(|\tilde{\sigma}|, \mu') \subseteq \text{Des}(\tilde{\sigma}, \mu')$. On the other hand, if $|\tilde{\sigma}(s)| = |\tilde{\sigma}(\text{South}(s))|$, then s is not a descent in $|\tilde{\sigma}|$, but will be a descent in $\tilde{\sigma}$ if and only if $\tilde{\sigma}(s)$ is a negative letter. Hence if we set $t = 1$ in (A.30) we get

$$\text{(A.31)} \qquad J_\mu(Z; q, 1) = \sum_{\text{nonattacking fillings } \sigma \text{ of } \mu'} z^\sigma q^{\text{maj}(\sigma,\mu')}$$

$$\times \prod_{\substack{w \in \mu' \\ \sigma(w)=\sigma(\text{South}(w))}} (1 - q^{\text{leg}(w)+1}) \prod_{\substack{w \in \mu' \\ \sigma(w) \neq \sigma(\text{South}(w))}} (1 - 1),$$

where for example the $-q^{\text{leg}(w)+1}$ term corresponds to choosing the negative letter for $\tilde{\sigma}(w)$.

One easily checks that if u, v, w is a triple as in Remark A.2, and $|\tilde{\sigma}(u)| = |\tilde{\sigma}(w)|$, then u, v, w form a coinversion triple if and only if $\tilde{\sigma}(u)$ contains a negative letter. (Here we use the fact that $|\tilde{\sigma}(u)| \neq |\tilde{\sigma}(v)|$ by the nonattacking condition). On the other hand if $|\tilde{\sigma}(u)| \neq |\tilde{\sigma}(w)|$, then $|\tilde{\sigma}(u)|, |\tilde{\sigma}(w)|, |\tilde{\sigma}(v)|$ are all distinct, and form a coinversion triple in $\tilde{\sigma}$ if and only if they form such a triple in $|\tilde{\sigma}|$. Putting everything together, we get the following combinatorial formula for J_μ.

COROLLARY A.11.1. [**HHL05a**]

$$(A.32) \qquad J_\mu(Z; q, t) = \sum_{\substack{nonattacking\ fillings\ \sigma\ of\ \mu'}} z^\sigma q^{maj(\sigma, \mu')} t^{coinv(\sigma, \mu')}$$

$$\times \prod_{\substack{u \in \mu' \\ \sigma(u) = \sigma(South(u))}} (1 - q^{leg(u)+1} t^{arm(u)+1}) \prod_{\substack{u \in \mu' \\ \sigma(u) \neq \sigma(South(u))}} (1 - t),$$

where each square in the bottom row is included in the last product.

EXAMPLE A.12. Let $\mu = (3, 3, 1)$. Then for the nonattacking filling σ of μ' in Figure 5, maj $= 3$, coinv $= 3$, squares $(1, 1), (1, 2), (2, 1), (2, 3)$ and $(3, 1)$ each contribute a $(1 - t)$, square $(1, 3)$ contributes a $(1 - qt^2)$, and $(2, 2)$ contributes a $(1 - q^2 t)$. Thus the term in (A.32) corresponding to σ is

$$(A.33) \qquad x_1 x_2^3 x_3^2 x_4 q^3 t^3 (1 - qt^2)(1 - q^2 t)(1 - t)^5.$$

FIGURE 5. A nonattacking filling of $(3, 3, 1)'$.

The (integral form) Jack polynomials $J_\mu^{(\alpha)}(Z)$ can be obtained from the Macdonald J_μ by

$$(A.34) \qquad J_\mu^{(\alpha)}(Z) = \lim_{t \to 1} \frac{J_\mu(Z; t^\alpha, t)}{(1 - t)^{|\mu|}}.$$

If we set $q = t^\alpha$ in (A.32) and then divide by $(1 - t)^{|\mu|}$ and take the limit as $t \to 1$ we get the following result of Knop and Sahi [**KS97**].

$$(A.35)$$
$$J_\mu^{(\alpha)}(Z) = \sum_{\substack{nonattacking\ fillings\ \sigma\ of\ \mu'}} z^\sigma \prod_{\substack{u \in \mu' \\ \sigma(u) = \sigma(South(u))}} (\alpha(leg(u) + 1) + arm(u) + 1).$$

There is another formula for J_μ corresponding to involution IN and (A.17). First note that by applying the $t \to 1/t$, $X \to tZ$ case of (A.22) to the right-hand side of (A.28) we get

$$(A.36) \qquad J_\mu(Z; q, t) = \sum_{\tilde{\sigma}: \mu \to \mathcal{A}_\pm, primary} z^{|\tilde{\sigma}|} t^{nondes(\tilde{\sigma}, \mu)} q^{inv(\tilde{\sigma}, \mu)} (-t)^{neg(\tilde{\sigma})},$$

where the sum is over all primary super fillings of μ, and *nondes* $= n(\mu) -$ maj is the sum of leg $+ 1$ over all squares of μ which are *not* in Des$(\tilde{\sigma}, \mu)$ and also not in the bottom row of μ. We will again group together the 2^n super fillings $\tilde{\sigma}$ with fixed absolute value σ.

For a positive filling σ and $s \in \mu$, let

$$(\text{A.37}) \qquad \text{maj}_s(\sigma, \mu) = \begin{cases} \text{leg}(s) \text{ if North}(s) \in \text{Des}(\sigma, \mu) \\ 0 \text{ else} \end{cases}$$

$$(\text{A.38}) \qquad \text{nondes}_s(\sigma, \mu) = \begin{cases} \text{leg}(s) + 1 \text{ if } s \notin \text{Des}(\sigma, \mu) \text{ and South}(s) \in \mu \\ 0 \text{ else,} \end{cases}$$

so $\text{maj} = \sum_s \text{maj}_s$ and $\text{nondes} = \sum_s \text{nondes}_s$. Note that a square s with South$(s) \in \mu$ and $s \notin \text{Des}(\sigma)$ makes a nonzero contribution to $\text{nondes}(\tilde{\sigma}, \mu)$ if and only if $\tilde{\sigma}(s)$ is a positive letter. Similarly, if s makes a nonzero contribution to $\text{maj}_s(\sigma, \mu)$, then s will make this same nonzero contribution to $\text{nondes}(\tilde{\sigma}, \mu)$ if and only if $\tilde{\sigma}(s)$ is a negative letter. Thus letting $t = 1$ in (A.36) we get

$$(\text{A.39}) \qquad J_\mu(Z; q, 1) = \sum_{\sigma: \mu \to \mathcal{A}_+, \text{primary}} z^\sigma \times \prod_{s \in \mu} (t^{\text{nondes}_s(\sigma, \mu)} - t^{1 + \text{maj}_s(\sigma, \mu)}).$$

Given a triple u, v, w of μ with $w \in \mu$, we define the "middle square" of u, v, w, with respect to $\tilde{\sigma}$, to be the square containing the middle of the three numbers of the set $\{|\tilde{\sigma}|'(u), |\tilde{\sigma}|'(v), |\tilde{\sigma}|'(w)\}$. In other words, we replace all letters by their absolute value, then standardize (with respect to ordering (A.17)), then take the square containing the number which is neither the largest, nor the smallest, of the three. For squares $(1, 1), (1, 2), (3, 2)$ for the super filling on the left in Figure 3, square $(1,2)$ is the middle square. For squares $(1, 4), (1, 3), (2, 4)$ square $(1, 3)$ is the middle square. Extend this definition to triples u, v, w of μ with u, v in the bottom row by letting the middle square be u if $|\tilde{\sigma}|(u) \leq |\tilde{\sigma}|(v)$, otherwise let it be v (here, as usual, v is to the right of u). By checking the eight possibilities for choices of signs of the letters in u, v, w (or four possibilities if u, v are in the bottom row), one finds that if u, v, w form a coinversion triple in $|\tilde{\sigma}|$, then they form a coinversion triple in $\tilde{\sigma}$ if the middle square contains a positive letter, otherwise they form an inversion triple in $\tilde{\sigma}$. Similarly, if u, v, w form an inversion triple in $|\tilde{\sigma}|$, they form a coinversion triple in $\tilde{\sigma}$ if the middle square contains a negative letter, otherwise they form an inversion triple. Thus by defining $\text{coinv}_s(\sigma, \mu)$ to be the number of coinversion triples for which s is the middle square, and $\text{inv}_s(\sigma, \mu)$ to be the number of inversion triples for which s is the middle square, we get the following (hitherto unpublished) result of the author.

COROLLARY A.12.1.
(A.40)

$$J_\mu(Z; q, t) = \sum_{\substack{\sigma: \mu \to \mathcal{A}_+ \\ \text{primary}}} z^\sigma \prod_{s \in \mu} (q^{inv_s(\sigma, \mu)} t^{\text{nondes}_s(\sigma, \mu)} - q^{coinv_s(\sigma, \mu)} t^{1 + \text{maj}_s(\sigma, \mu)}).$$

Although the products in (A.40) are more complicated than those in (A.32), one advantage this formula has is that the q and t-weights are invariant under standardization, hence we can express it as

$$(\text{A.41}) \qquad J_\mu(Z; q, t) = \sum_{\substack{\beta \in S_n \\ \text{primary}}} Q_{n, \text{Des}(\beta^{-1})}(Z)$$

$$\times \prod_{s \in \mu} (q^{inv_s(\beta, \mu)} t^{\text{nondes}_s(\beta, \mu)} - q^{coinv_s(\beta, \mu)} t^{1 + \text{maj}_s(\beta, \mu)}),$$

which gives an expansion of J_μ into Gessel's fundamental quasi-symmetric functions. The sum is over all standard fillings (permutations) which are also primary fillings, although the equation is still true if we extend the sum to all permutations, since if β is a standard filling which is not primary, with pivotal square s, then the term $\left(q^{inv_s(\beta,\mu)} t^{nondes_s(\beta,\mu)} - q^{coinv_s(\beta,\mu)} t^{1+maj_s(\beta,\mu)} \right)$ is zero.

Schur Coefficients. Since by (A.14) $\tilde{H}_\mu(X;q,t)$ is a positive sum of LLT polynomials, Grojnowski and Haiman's result [**GH06**], that LLT polynomials are Schur positive, gives a new proof that $\tilde{K}_{\lambda,\mu}(q,t) \in \mathbb{N}[q,t]$. In fact, we also get a natural decomposition of $\tilde{K}_{\lambda,\mu}(q,t)$ into "LLT components". This result is geometric though, and doesn't yield combinatorial formulas for these Schur coefficients. Sami Assaf [**Ass07b**] has introduced the concept of a *dual equivalence graph* and, utilizing their properties, has found a way to prove Schur positivity for LLT products of any three skew shapes. This gives the first combinatorial proof of Schur positivity for the $\tilde{H}_\mu$ when μ has three columns. As mentioned in the preface and in Chapter 6, her recent preprint [**Ass07a**] extends this method to prove Schur positivity of general LLT polynomials, and hence Macdonald polynomials, combinatorially. At this time her construction does not easily lead to a elegant closed form expression for the Schur coefficients for shapes with more than two columns, but it seems refinements of her method may well do so.

In this section we indicate how nice combinatorial formulas for the $\tilde{K}_{\lambda,\mu}(q,t)$ can be obtained from Theorem A.3 when μ is either a hook shape or has two columns.

The Hook Case. Given a filling σ of $\mu = (n-k,1^k)$, by applying the map coϕ from Exercise 1.5 we see Theorem A.3 implies

$$(A.42) \qquad \tilde{H}_{(n-k,1^k)}(X;q,t) = \sum_\sigma x^\sigma t^{maj(\sigma_1,\dots,\sigma_k)} q^{comaj(\sigma_k,\dots,\sigma_n)}.$$

Theorem 1.23 implies that the q and t powers in (A.42) depend only on the recording tableau $Q_{read(\sigma)}$. Since the x^σ corresponding to the $P_{read(\sigma)}$ for a fixed shape λ generate the Schur function s_λ, we have the following variant of Stembridge's formula (2.27).

$$(A.43) \qquad \tilde{K}_{\lambda,(n-k,1^k)}(q,t) = \sum_{T \in SYT(\lambda)} t^{maj(T;\mu')} q^{comaj(T;rev(\mu))}.$$

Butler [**But94**, pp. 109-110] gives a bijective map B between SYT of fixed shape λ with the property that i is a descent of T if and only if $n-i+1$ is a descent of $B(T)$. Applying this to the T in the sum above gives

$$(A.44) \qquad \tilde{K}_{(n-k,1^k)}(q,t) = \sum_{B \in SYT(\lambda)} t^{maj(B;\mu)} q^{comaj(B;rev(\mu'))},$$

which is equivalent to Stembridge's formula after applying (2.30).

REMARK A.13. The analysis above gives a solution to Problem A.6 for hook shapes at the level of Schur functions.

The Two-Column Case. A final segment of a word is the last k letters of the word, for some k. We say a filling σ is a *Yamanouchi filling* if in any final segment of read(σ), there are at least as many i's as $i+1$'s, for all $i \geq 1$. In [**HHL05a**] the following result is proved.

THEOREM A.14. *For any partition μ with $\mu_1 \leq 2$,*

$$(A.45) \qquad \tilde{K}_{\lambda,\mu}(q,t) = \sum_{\sigma \, Yamanouchi} t^{maj(\sigma,\mu)} q^{inv(\sigma,\mu)},$$

where the sum is over all Yamanouchi fillings of μ.

The proof of Theorem A.14 involves a combinatorial construction which groups together fillings which have the same maj and inv statistics. This is carried out with the aid of *crystal graphs*, which occur in the representation theory of Lie algebras. We should mention that in both (A.45) and (A.44), if we restrict the sum to those fillings with a given descent set, we get the Schur decomposition for the corresponding LLT polynomial.

If, in (A.45), we relax the condition that μ has at most two columns, then the equation no longer holds. It is an open problem to find a way of modifying the concept of a Yamanouchi filling so that (A.45) is true more generally. A specific conjecture, when μ has three columns, for the $\tilde{K}_{\lambda,\mu}(q,t)$, of the special form (2.21), is given in [**Hag04a**], although it does not seem to be translatable into a sum over fillings as in (A.45).

Understanding the Schur coefficients of LLT and Macdonald polynomials is also connected to a fascinating group of problems associated to a family of symmetric functions $s_\lambda^{(k)}(X;t)$ introduced by Lapointe, Lascoux and Morse called k-Schur functions [**LLM03**]. For $k = \infty$ these functions reduce to the ordinary Schur function, but in general they depend on a parameter t. They conjecture that when Macdonald polynomials are expanded into k-Schur functions, for k beyond a certain range the coefficients are in $\mathbb{N}[q,t]$, and furthermore, when expanding the k-Schur into Schur functions the coefficients are in $\mathbb{N}[t]$. Hence this conjecture refines Macdonald positivity. In a series of subsequent papers, Lapointe and Morse have introduced a number of other combinatorial conjectures involving the k-Schur [**LM03a**],[**LM03c**],[**LM04**],[**LM05**]. They have shown that the k-Schur appears to have several equivalent definitions, and for each of these definitions they prove a different subset of these other conjectures. Interest in the k-Schur has deepened in the last few years as Lapointe and Morse have shown that when $t = 1$, the k-Schur have applications to the study of the quantum cohomolgy of the Grassmannian. [**LM**]. Building on conjectures of Shimozono and joint work of Lam, Lapointe, Morse and Shimozono, [**LLMS06**], Lam [**Lam**] has shown that the the k-Schur and variants of the k-Schur, known as dual k-Schur, are polynomial representatives for the Schubert classes in homology and cohomology, respectively, of the affine Grassmannian. Recently it has been informally conjectured by some researchers in this area that, when expanding LLT polynomials of total "bandwidth" k (i.e., all of whose skew shapes fit within a span of k diagonals, so for example the LLT polynomial corresponding to Figure 2 would have bandwidth 4) into the k-Schur functions $s_\lambda^{(k)}(X;q)$, the coefficients are in $\mathbb{N}[q]$. Hence we can expand the $\tilde{H}_\mu(X;q,t)$ positively in terms of LLT polynomials, and (conjecturally) LLT 's positively in terms of k-Schur and (again conjecturally) k-Schur positively in terms of Schur.

Nonsymmetric Macdonald Polynomials

In 1995 Macdonald [**Mac96**] introduced polynomials $E'_\alpha(X;q,t)$ which form a basis for the polynomial ring $\mathbb{Q}[x_1,\ldots,x_n](q,t)$, and in many ways are analogous to the $P_\lambda(X;q,t)$. Further development of the theory was made by Cherednik

[**Che95**], Sahi [**Sah98**], Knop [**Kno97a**], Ion [**Ion03**], and others. Both the E'_α and the P_λ have versions for any "affine" root system, and are both orthogonal families of polynomials with respect to a certain scalar product.

We will be working with the type A_{n-1} case, where there is a special structure which allows us to assume α is a composition, i.e. $\alpha \in N^n$. Given $\alpha \in N^n$, let α' denote the "transpose graph" of α, consisting of the squares

$$(A.46) \qquad \alpha' = \{(i,j), \quad 1 \leq i \leq n, \quad 1 \leq j \leq \alpha_i)\}$$

Furthermore let $dg(\alpha')$ denote the augmented diagram obtained by adjoining the "basement" row of n squares $\{(j,0), 1 \leq j \leq n\}$ below α'. Given $s \in \alpha'$, we let leg(s) be the number of squares of α' above s and in the same column of s. Define Arm(s) to be the set of squares of $dg(\alpha')$ which are either to the right and in the same row as s, and also in a column not taller than the column containing s, or to the left and in the row below the row containing s, and in a column strictly shorter than the column containing s. Then set arm(s) = |Arm(s)|. For example, for $\alpha = (1,0,3,2,3,0,0,0,0)$, the leg lengths of the squares of $(1,0,3,2,3,0,0,0,0)'$ are listed on the left in Figure 6 and the arm lengths on the right. Note that if α is a partition μ, the leg and arm definitions agree with those previously given for μ'. We will assume throughout this section that if $\alpha \in N^n$, then $\alpha_1 + \ldots + \alpha_n \leq n$.

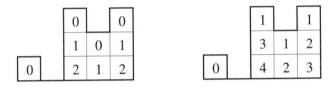

FIGURE 6. The leg lengths (on the left) and the arm lengths (on the right) for $(1,0,3,2,3,0,0,0,0)'$.

Let

$$(A.47) \qquad E_\alpha(x_1, \ldots, x_n; q, t) = E'_{\alpha_n, \ldots, \alpha_1}(x_n, \ldots, x_1; 1/q, 1/t).$$

This modified version of the E'_α occurs in a paper of Marshall [**Mar99**], who made a special study of of the E_α in the type A_{n-1} case, showing among other things they satisfy a version of Selberg's integral. Assume that $t = q^k$ for $k \in \mathbb{N}$. Then given two polynomials $f(x_1, \ldots, x_n; q, t)$ and $g(x_1, \ldots, x_n; q, t)$ whose coefficients depend on q and t, define a scalar product

$$(A.48) \qquad \langle f, g \rangle'_{q,t} = \mathrm{CT} f(x_1, \ldots, x_n; q, t) g(1/x_1, \ldots, 1/x_n; 1/q, 1/t)$$

$$\times \prod_{1 \leq i < j \leq n} (\frac{x_i}{x_j}; q)_k (q\frac{x_j}{x_i}; q)_k,$$

where CT means "take the constant term in". Then (when $t = q^k$) both the (type A_{n-1}) E_α and the P_λ are orthogonal with respect to $\langle, \rangle'_{q,t}$. (This fact has an extension to general affine root systems, while it is not known whether the orthogonality of the P_λ with respect to the inner product in (2.3) has a version for other root systems.)

Define the "integral form" nonsymmetric Macdonald polynomial $\mathcal{E}_\alpha$ as

$$(A.49) \qquad \mathcal{E}_\alpha(X; q, t) = E_\alpha(X; q, t) \prod_{s \in \alpha'} (1 - q^{\mathrm{leg}(s)+1} t^{\mathrm{arm}(s)+1}).$$

Theorem A.15 below describes a combinatorial formula for $\mathcal{E}_\alpha(X; q, t)$ which, in short, is the same as (A.32) after extending the definitions of arm, leg, coinv, maj to composition diagrams appropriately, and changing the basement to $\sigma(j, 0) = j$.

Given $\alpha \in \mathbb{N}^n$, we define a triple of squares of α' to be three squares u, v, w, with $u \in \alpha'$, $w = \text{South}(u)$, and $v \in \text{Arm}(u)$. Note that v, w need not be in α', i.e. they could be in the basement. A filling σ of α' is an assignment of integers from the set $\{1, \ldots, n\}$ to the squares of α'. As before, we let the reading word $\text{read}(\sigma)$ be the word obtained by reading across rows, left to right, top to bottom. The standardization of a filling σ is the filling whose reading word is the standardization of $\text{read}(\sigma)$. We determine the orientation of a triple by starting at the smallest and going in a circular motion to the largest, where if two entries of a triple have equal σ-values then the one that occurs earlier in the reading word is viewed as being smaller.

We say a square $s \in \alpha'$ attacks all squares to its right in its row and all squares of $\text{dg}(\alpha')$ to its left in the row below. Call a filling nonattacking if there are no pairs of squares (s, u) with $s \in \alpha'$, s attacks u, and $\sigma(s) = \sigma(u)$. Note that, since $\sigma(j, 0) = j$, in any nonattacking filling with s of the form $(k, 1)$, we must have $\sigma(s) \geq k$. Figure 7 gives a nonattacking filling of $(1, 0, 3, 2, 3, 0, 0, 0, 0)'$.

As before we let $\text{South}(s)$ denote the square of $\text{dg}(\alpha')$ immediately below s, and let $\text{maj}(\sigma, \alpha')$ denote the sum, over all squares $s \in \alpha'$ satisfying $\sigma(s) > \sigma(\text{South}(s))$, of $\text{leg}(s) + 1$. A triple of α' is three squares with $u \in \alpha'$, $v \in \text{Arm}(u)$, and $w = \text{South}(u)$. We say such a triple is a coinversion triple if either v is in a column to the right of u, and u, v, w has a clockwise orientation, or v is in a column to the left of u, and u, v, w has a counterclockwise orientation. Let $\text{coinv}(\sigma, \alpha')$ denote the number of coinversion triples of σ. For example, the filling in Figure 7 has coinversion triples

(A.50) $\{[(3, 2), (3, 1), (4, 2)], [(3, 2), (3, 1), (5, 2)], [(3, 2), (3, 1), (1, 1)],$
$[(4, 2), (4, 1), (1, 1)], [(5, 1), (5, 0), (1, 0)], [(5, 1), (5, 0), (2, 0)], [(5, 1), (5, 0), (4, 0)]\}.$

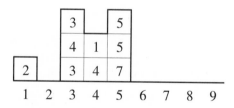

FIGURE 7. A nonattacking filling of $(1, 0, 3, 2, 3, 0, 0, 0, 0)'$.

THEOREM A.15. [**HHL**] *For* $\alpha \in \mathbb{N}^n$, $\sum_i \alpha_i \leq n$,

(A.51) $\mathcal{E}_\alpha(x_1, \ldots, x_n; q, t) = \displaystyle\sum_{\substack{\text{nonattacking fillings } \sigma \text{ of } \alpha'}} x^\sigma q^{\text{maj}(\sigma, \alpha')} t^{\text{coinv}(\sigma, \alpha')}$

(A.52) $\times \displaystyle\prod_{\substack{u \in \mu' \\ \sigma(u) = \sigma(\text{South}(u))}} (1 - q^{\text{leg}(u)+1} t^{\text{arm}(u)+1}) \prod_{\substack{u \in \mu' \\ \sigma(u) \neq \sigma(\text{South}(u))}} (1 - t),$

where $x^\sigma = \prod_{s \in \alpha'} x_{\sigma(s)}$.

EXAMPLE A.16. By (A.50) the nonattacking filling in Figure 7 has coinv = 7. There are descents at squares $(1,1)$, $(3,2)$ and $(5,1)$, with maj-values 1, 2 and 1, respectively. The squares $(3,1)$, $(4,1)$ and $(5,3)$ satisfy the condition $\sigma(u) = \sigma(\mathrm{South}(u))$ and contribute factors $(1 - q^3t^5)$, $(1 - q^2t^3)$ and $(1 - qt^2)$, respectively. Hence the total weight associated to this filling in (A.51) is

$$(A.53) \qquad x_1x_2x_3^2x_4^2x_5^2x_7q^4t^7(1 - q^3t^5)(1 - q^2t^3)(1 - qt^2)(1 - t)^6.$$

REMARK A.17. Cherednik has shown that the E'_α satisfy two recurrence relations (known as intertwiners) which uniquely determine them. In general these are difficult to interpret combinatorially, but in the type A case Knop [**Kno97a**] used the cyclic symmetry of the affine root system $\hat{A}_{n-1}$ to transform one of these into a simpler relation. Expressed in terms of the E_α this says that if $\alpha_n > 0$,

$$(A.54) \qquad E_\alpha(X; q, t) = x_n q^{\alpha_n - 1} E_{\alpha_n - 1, \alpha_1, \alpha_2, \dots, \alpha_{n-1}}(qx_n, x_1, x_2, \dots, x_{n-1}).$$

This relation can be interpreted bijectively (Exercise A.18). In [**HHL**] it is shown that the other intertwiner relation is also satisfied by the E_α, a key point of which is to show that if $\alpha_i = \alpha_{i+1}$, then $E_\alpha(X; q, t)$ is symmetric in x_i, x_{i+1}.

EXERCISE A.18. Assuming $\alpha_n > 0$, give a bijective proof that (A.54) holds.

REMARK A.19. For any $\alpha \in \mathbb{N}^n$, let α^+ denote the partition obtained by rearranging the parts of α into nonincreasing order. It is well-known that the $P_\lambda(X; q, t)$ can be expressed as a linear combination of those E'_α for which $\alpha^+ = \lambda$. In terms of the E_α, this identity takes the following form [**Mar99**]

$$(A.55) \quad P_\lambda(X; q, t) = \prod_{s\in\lambda'}(1 - q^{\mathrm{leg}(s)+1}t^{\mathrm{arm}(s)}) \sum_{\alpha:\alpha^+=\lambda} \frac{E_\alpha(X; q, t)}{\prod_{s\in\alpha'}(1 - q^{\mathrm{leg}(s)+1}t^{\mathrm{arm}(s)})}.$$

If we set $q = t = 0$ in (A.55), then by Remark 2.3 we have the identity

$$(A.56) \qquad s_\lambda(X) = \sum_{\alpha:\alpha^+=\lambda} NS_\alpha(X),$$

where $NS_\alpha(X)$ is the sum of x^σ over all fillings σ of $\mathrm{dg}(\alpha')$ with no descents and no coinversion triples. Sarah Mason has proved this identity bijectively by developing a generalization of the RSK algorithm [**Mas06**], [**Mas**].

REMARK A.20. Let μ be a partition, and $\alpha \in \mathbb{N}^n$ with $(\alpha^+)' = \mu$, that is, a diagram obtained by permuting the columns of μ. If we let $\sigma(j, 0) = \infty$, it is an easy exercise to show the involutions IM and IN from the proof of Theorem A.3 hold for fillings of α'. It follows that formula (A.4) for $\tilde{H}_\mu(X; q, t)$ and (A.41) for $J_\mu(Z; q, t)$ both hold if, instead of summing over fillings of μ, we sum over fillings of α', using the definitions of arm, leg, etc. given earlier in this section. Similarly, formula (A.32) for $J_\mu(X; q, t)$ holds if we replace nonattacking fillings of μ' by nonattacking fillings of any permutation of the columns of μ'.

REMARK A.21. Knop and Sahi obtained a combinatorial formula for the nonsymmetric Jack polynomial, which is a limiting case of (A.51) in the same way that (A.35) is a limiting case of (A.32). In fact, it was contrasting their formula for the nonsymmetric Jack polynomials with (A.32) which led to (A.51).

The Genesis of the Macdonald Statistics

In this section we outline the empirical steps which led the author to the monomial statistics for $\tilde{H}_\mu$. The fact that the sum of $x^\sigma t^{\mathrm{maj}(\sigma,\mu)}$ generates $\tilde{H}_\mu(X;1,t)$ can be derived from Theorem A.3 by applying the RSK algorithm to the "column reading word" $\mathrm{col}(\sigma)$ obtained by reading the entries of σ down columns, starting with the leftmost column and working to the right. By Theorem 1.23 the descent set of σ can be read off from the descent set of $Q_{\mathrm{col}(\sigma)}$. See also [**Mac95**, p. 365]. On the other hand, given the connections between Macdonald polynomials and diagonal harmonics, in light of the shuffle conjecture one might suspect that the correct q-statistic to match with maj involves dinv in some way.

By definition,

$$(A.57) \qquad \tilde{H}_\mu[Z;q,t]\big|_{t^{n(\mu)}} = \sum_\lambda s_\lambda t^{n(\mu)} K_{\lambda,\mu}(q,1/t)\big|_{t^{n(\mu)}}$$

$$(A.58) \qquad = \sum_\lambda s_\lambda K_{\lambda,\mu}(q,0)$$

$$(A.59) \qquad = \sum_\lambda s_\lambda K_{\lambda',\mu'}(0,q),$$

by (2.29). Now Theorem 6.8 shows that if $\pi = \pi(\mu')$ is the balanced Dyck path whose consecutive bounce steps are the parts of μ' in reverse order, then

$$(A.60) \qquad \mathcal{F}_\pi(X;q) = q^{\mathrm{mindinv}(\pi)} \sum_\lambda s_\lambda(X) K_{\lambda',\mu'}(q).$$

Comparing (A.60) and (A.59) we are led naturally to consider $\mathcal{F}_\pi(X;q)$ in connection with $\tilde{H}_\mu(X;q,t)$.

If we start with a word parking function β for π, push all columns down so their bottom squares all align in row, remove all empty columns, and finally reverse the order of the columns, we get a filling $\sigma(\beta)$ of μ. See Figure 8 for an example.

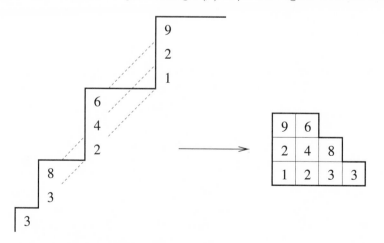

FIGURE 8. Transforming a word parking function for the Dyck path $\pi(1,2,3,3)$ into a filling of $(3,3,2,1)' = (4,3,2)$.

One checks that $\mathrm{Inv}(\sigma(\beta),\mu) = \mathrm{dinv}(\beta)$. Since word parking functions are decreasing down columns, $\mathrm{maj}(\sigma(\beta)) = n(\mu)$. Comparing (A.60) and (A.59) we now see that

$$(A.61) \qquad \tilde{H}_\mu(X;q,t)|_{t^{n(\mu)}} = q^{-\mathrm{mindinv}(\pi)} \sum_\sigma x^\sigma q^{\mathrm{Inv}(\sigma,\mu)},$$

where the sum is over all fillings σ of μ which are decreasing down columns.

After further experimentation the author noticed that the sum of $x^\sigma q^{\mathrm{Inv}(\sigma,\mu)}$ over all fillings with no descents seemed to generate $\tilde{H}_\mu(X;q,0)$. Thus we can generate the terms in $\tilde{H}_\mu$ corresponding to descents everywhere by subtracting mindinv(π) from Inv and the terms corresponding to no descents by subtracting nothing from Inv. This led to the the hypothesis that the statistic $\mathrm{Inv}(\sigma,\mu)$ is an upper bound for the correct q-power, in the sense that it always seemed possible to subtract something nonnegative, possibly depending on the descents in some way, from Inv and correctly generate $\tilde{H}_\mu(X;q,t)$. After a period of trial and error the author finally realized that mindinv could be viewed as the sum, over all squares s not in the bottom row, of $\mathrm{arm}(s)$. This set of squares is exactly the set where descents occur in the $\sigma(\beta)$ fillings, and a Maple program verified that subtracting $\mathrm{arm}(s)$, over all $s \in \mathrm{Des}(\sigma,\mu)$, from Inv did indeed give the correct q-statistic to match with $x^\sigma t^{\mathrm{maj}}$.

REMARK A.22. We saw in Chapter 2 that results in Macdonald polynomials are often inspired by corresponding results involving Jack polynomials. This didn't happen with Knop and Sahi's formula (A.35) though. Certainly the

$$z^\sigma \prod (1 - q^{\mathrm{leg}(u)+1} t^{\mathrm{arm}(u)+1}) \prod (1-t)$$

portion of (A.32) could have been easily deduced from (A.35), but the study of the combinatorics of the space of diagonal harmonics was crucial to discovering the $q^{\mathrm{maj}(\sigma,\mu')} t^{\mathrm{coinv}(\sigma,\mu')}$ term.

APPENDIX B

The Loehr-Warrington Conjecture

The Conjecture

In this appendix we discuss a recent conjecture of Loehr and Warrington [**LW**], which gives a combinatorial formula for the expansion of $\nabla s_\lambda(X)$ into monomials, for any partition λ. If $\lambda = 1^n$, their conjecture reduces to Conjecture 6.1.

Given $\lambda \in \mathrm{Par}(n)$, we fill λ with border strips by repeatedly removing the entire northeast border of λ, as in Figure 1. We let $n_j = n_j(\lambda)$, $0 \leq j \leq \lambda_1 - 1$ denote the length of the border strip which starts in square $(\lambda_1 - j, 0)$, if a border strip does start there, and 0 otherwise. In Figure 1, we have

(B.1) $$(n_0, n_1, n_2, n_3, n_4) = (10, 7, 0, 3, 1).$$

(The dot in square $(1, 1)$ indicates the length of that border strip is 1.)

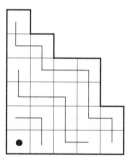

FIGURE 1. The decomposition of the partition $(5, 5, 4, 4, 2, 1)$ into border strips.

Define $\mathrm{spin}(\lambda)$ to be the total number of times some border strip of λ crosses a horizontal boundary of a unit square of λ. In Figure 1, spin $= 9$ as the border strips of lengths $(10, 7, 3, 1)$ contribute $(4, 3, 1, 0)$ to spin, respectively. Let

(B.2) $$\mathrm{sgn}(\lambda) = (-1)^{\mathrm{spin}(\lambda)},$$

and call

(B.3) $$\mathrm{adj}(\lambda) = \sum_{j : n_j > 0} (\lambda_1 - 1 - j)$$

the *dinv adjustment* $(5 + 3 + 1 + 0 = 9$ in Figure 1$)$.

A λ-family π is a sequence $(\pi_0, \pi_1, \ldots, \pi_{\lambda_1 - 1})$ where

(1) π_j is a lattice path consisting of unit North $N(0, 1)$ and East $E(1, 0)$ steps, starting at (j, j) and ending at $(j + n_j, j + n_j)$, which never goes below the diagonal $x = y$.

141

(2) No two paths π_j, π_k ever cross, and π_j never touches a diagonal point (k, k) for $j < k \leq \lambda_1 - 1$.

(3) Two different paths π_j, π_k can share N steps but not E steps.

Note that paths of length 0, corresponding to those j with $n_j = 0$, are relevant in that other paths must avoid those (j, j) points, and that a (1^n)-family reduces to an ordinary Dyck path. Figure 2 gives an example of a $(5, 5, 4, 4, 2, 1)$-family of paths. Here the path of length 0 is denoted by a black dot, and the nonzero paths are alternately drawn with solid and dotted lines to clearly distinguish them from one another.

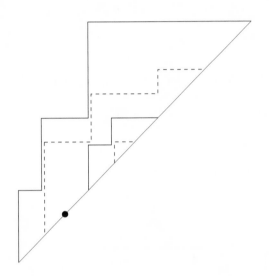

FIGURE 2. A $(5, 5, 4, 4, 2, 1)$-family of Dyck paths.

Let $a_i^{(j)}$ denote the number of squares in the ith row of path π_j (as defined below (1.14)) for $j + 1 \leq i \leq j + n_j$, with $a_i^{(j)}$ undefined for other values of i. We call this array of values the *Area array* of the λ-family π, denoted Area(π). For the family in Figure 2, we have Area array

$$\begin{pmatrix} a^{(0)} : 0 & 1 & 2 & 2 & 3 & 4 & 3 & 4 & 5 & 6 \\ a^{(1)} : - & 0 & 1 & 2 & 3 & 2 & 3 & 1 & - & - \\ a^{(2)} : - & - & - & - & - & - & - & - & - & - \\ a^{(3)} : - & - & - & 0 & 1 & 1 & - & - & - & - \\ a^{(4)} : - & - & - & - & 0 & - & - & - & - & - \end{pmatrix}$$

where undefined values are indicated by $-$'s. Denote the set of all Area arrays of λ-families of paths NDP_λ (NDP stands for "Nested Dyck Paths"), and for $A \in NDP_\lambda$, let

$$(B.4) \qquad \qquad \text{area}(A) = \sum_{i,j} a_i^{(j)},$$

where the sum is over all defined elements of A (so area $= 30 + 12 + 0 + 2 + 0 = 44$ for the above array).

DEFINITION B.1. A *Labeling* for a λ-family is an array $\{r_i^{(j)}\}$, where $r_i^{(j)}$ is a positive integer we associate with the $i-j$th N step of π_j. The $r_i^{(j)}$ are defined for exactly the same values of i, j that the $a_i^{(j)}$ are. In addition, we require

(1) If $a_{i+1}^{(j)} = a_i^{(j)} + 1$, then $r_i^{(j)} < r_{i+1}^{(j)}$. (decrease down columns)
(2) For $j < k$, if $a_p^{(j)}$ and $a_{p-1}^{(k)}$ are defined with $a_p^{(j)} = a_{p-1}^{(k)} + 1$, then $r_p^{(j)} \leq r_{p-1}^{(k)}$.

Condition (2) above implies that for a given column, no larger label in the row directly above is associated to a lower-indexed path. In Figure 3, the labels for π_0 are drawn just to the left of the N steps, and just to the right for the other paths. This corresponds to Labeling array

$$
\begin{pmatrix}
r^{(0)} : 2 & 4 & 6 & 1 & 3 & 6 & 5 & 7 & 9 & 10 \\
r^{(1)} : - & 2 & 4 & 5 & 6 & 6 & 7 & 10 & - & - \\
r^{(2)} : - & - & - & - & - & - & - & - & - & - \\
r^{(3)} : - & - & - & 1 & 8 & 10 & - & - & - & - \\
r^{(4)} : - & - & - & - & 11 & - & - & - & - & -
\end{pmatrix}
$$

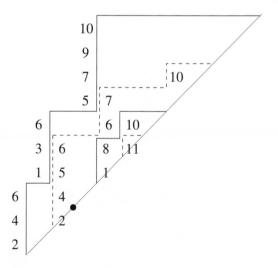

FIGURE 3. A Labeling for a $(5, 5, 4, 4, 2, 1)$-family.

We let $LNDP_\lambda$ denote the set of pairs (A, R), where A is an Area array for a λ-family π, and R is a Labeling of π. For $(A, R) \in LNDP_\lambda$, we let

(B.5) $\qquad \text{dinv}(A, R) = \text{adj}(\lambda) + \sum_{\substack{u,v \\ b \leq c}} \chi(a_b^{(u)} - a_c^{(v)} = 1)\chi(r_b^{(u)} > r_c^{(v)})$

$$+ \sum_{\substack{u,v \\ b < c \text{ or } (b = c \text{ and } u < v)}} \chi(a_b^{(u)} - a_c^{(v)} = 0)\chi(r_b^{(u)} < r_c^{(v)}),$$

recalling that for any logical statement S, $\chi(S) = 1$ if S is true, and 0 if S is false. Furthermore, set

$$(B.6) \qquad\qquad \text{area}(A) = \sum_{i,j} a_i^{(j)},$$

where the sum is over all defined elements of A (so $\text{area}(A) = 30 + 12 + 0 + 2 + 0 = 44$ for the above array).

We are now ready to state the Loehr-Warrington conjecture.

CONJECTURE B.2. *Given* $\lambda \in Par$,

$$(B.7) \qquad\qquad \nabla s_\lambda = sgn(\lambda) \sum_{(A,R) \in LNDP_\lambda} t^{area(A)} q^{dinv(A,R)} x_R,$$

where $x_R = \prod_{i,j} x_{r_i^{(j)}}$.

The reader can check that for $\lambda = 1^n$, the definition of dinv above reduces to the one for word parking functions in Chapter 6, and Conjecture B.2 becomes equivalent to Conjecture 6.1.

REMARK B.3. The matrix which expresses the monomial symmetric functions m_λ in terms of the Schur functions s_λ is known as the inverse Kostka matrix. It has integer entries, but not all of the same sign. Eğecioğlu and Remmel [**ER90**] have given a combinatorial interpretation of the entries of this matrix, as a signed sum over certain combinatorial objects. Loehr and Warrington use these and other more recent results along these lines to prove the $q = 1$ case of Conjecture B.2. They also note that Conjecture B.2 could potentially be proved bijectively, without recourse to plethystic identities involving rational functions as in Chapter 7. To see why, note that

$$(B.8) \qquad\qquad \nabla \tilde{H}_\mu = T_\mu \tilde{H}_\mu$$

$$(B.9) \qquad\qquad = T_\mu \sum_\lambda D_{\lambda,\mu}(q,t) m_\lambda,$$

where $D_{\lambda,\mu}(q,t)$ is the sum of the q,t-monomials in (A.4) with x-weight x^λ. On the other hand,

$$(B.10) \qquad\qquad \nabla \tilde{H}_\mu = \sum_\lambda D_{\lambda,\mu}(q,t) \nabla m_\lambda$$

$$(B.11) \qquad\qquad = \sum_\lambda D_{\lambda,\mu}(q,t) \sum_\beta K_{\lambda,\beta}^{-1} \nabla s_\beta,$$

where $K_{\lambda,\beta}^{-1}$ is an entry in the inverse Kostka matrix. Eq. (B.11) uniquely determines the ∇s_λ, so if one shows that (B.11) holds with ∇s_λ replaced by the right-hand side of (B.7), then Conjecture B.2 follows.

Expansion into LLT polynomials

In this section we present an argument of Loehr and Warrington which shows that the right-hand side of (B.7) is a positive sum of LLT polynomials, and is hence a Schur-positive symmetric function. Starting with a pair $(A, R) \in LNDP_\lambda$, we first divide the labels $r_i^{(j)}$ into multisets, where two labels (possibly in different π_j) are in the same multiset if they are in the same column, and between their corresponding North steps there are no unit $(0, 1)$ segments of the underlying grid which are not in any π_k. For example, for the pair (A, R) in Figure 3, there are no cases where two North steps in the same column are separated by an empty $(0, 1)$ segment, and reading down the columns from left to right we get multisets

(B.12) $\{2, 4, 6\}, \{1, 2, 3, 4, 5, 6, 6\}, \{1, 5, 6, 7, 7, 8, 9, 10\}, \{10, 11\}, \{10\}.$

In Figure 4 however, the labels in column 4 break into multisets $\{9, 10\}$ and $\{1, 6, 7, 8\}$ with the North steps corresponding to the two multisets separated by the unit segment from $(3, 7)$ to $(3, 8)$.

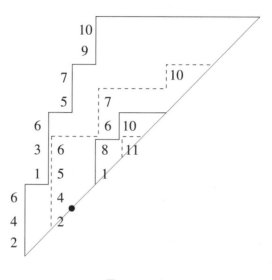

FIGURE 4

Number the multisets as we encounter them reading down columns, left to right, $L_1, L_2, \ldots, L_p$. Say there are b paths $\pi_{m-b+1}, \ldots, \pi_{m-1}, \pi_m$ whose labels contribute to L_k. We construct an SSYT T_k whose ith column from the right contains those labels from π_{m+i-1} in L_k. This will be part of a term $T = T(A, R) = (T_1, \ldots, T_p)$ in an LLT polynomial as in Figure 5. Note that the label from L_k which is closest to the line $x = y$ must be from π_m. This label, and the other labels in L_k from π_m, form the rightmost column of T_k, and are on the same diagonal in T that they are in (A, R). For $i > 1$, to form the ith column from the right of T_k, take the labels from π_{m-i+1} and shift them i columns to the left and i rows downward in relation to where they are in (A, R). Note that this is a diagonal-preserving shift.

The fact that no two paths share an E step and no path intersects the beginning (i, i) of any other path implies that, for fixed j, in (A, R) the row containing the smallest label in L_k from π_{m-j+1} is at least one row above the row containing the smallest label in L_k from π_{m-j+2}, and the row containing the largest label in L_k from π_{m-j+1} is at least one row above the row containing the largest label in L_k from π_{m-j+2}. It follows that the shape of T_k is a skew shape, denoted $\beta^{(k)}/\nu^{(k)}$. It is now easy to see that for a fixed λ-family π, the construction above is a bijection between terms (A, R) in $LNDP_\lambda$ and SSYT $T(A, R)$ in the corresponding LLT product of skew shapes $(\beta^{(1)}(A)/\nu^{(1)}(A), \ldots, \beta^{(p)}(A)/\nu^{(p)}(A))$. This bijection trivially fixes

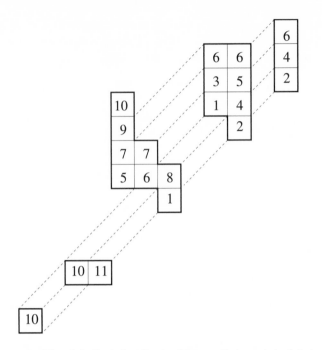

FIGURE 5. The labelled family in Figure 3 translated into the corresponding term in an LLT polynomial.

the x-weight. By Exercise B.4 below, we have

$$
\text{(B.13)} \qquad \sum_{(A,R)\in LNDP_\lambda} t^{\text{area}(A)} q^{\text{dinv}(A,R)} x_R = q^{\text{adj}(\lambda)} \sum_{A\in NDP_\lambda} t^{\text{area}(\pi)} q^{\text{pow}(\beta/\nu)}
$$

$$
\times \sum_{T\in SSYT(\beta/\nu(A))} q^{\text{inv}(T)} x_T,
$$

for a certain integer $\text{pow}(\beta/\nu)$. Since the inner sum on the right-hand side above is an LLT polynomial, we have proved the claim that the right-hand side of Conjecture B.7 is a positive sum of LLT polynomials.

EXERCISE B.4. For a given skew shape π/ζ, let $\text{ndiag}_p(\pi/\zeta)$ equal the number of squares of π/ζ whose coordinates (i,j) satisfy $j - i = p$. Given $(A,R) \in LNDP_\lambda$, show that

$$
\text{(B.14)} \qquad \text{dinv}(A,R) = \text{adj}(\lambda) + \text{pairdiag}(T) + \text{inv}(T(A,R)),
$$

where inv is the LLT inversion statistic from (6.7) and

$$
\text{(B.15)} \qquad \text{pairdiag}(T) = \sum_{k,p} \binom{\text{ndiag}_p(\beta^{(k)}/\nu^{(k)})}{2}.
$$

REMARK B.5. Loehr and Warrington have another conjecture [**LW07**] which gives a combinatorial interpretation for the monomial expansion of ∇p_n in terms of labelled lattice paths inside a square. Can and Loehr [**CL06**] have proved the sign-character restriction of this conjecture, which involves a bounce statistic for lattice paths inside a square. Their proof illustrates some refined properties of the bounce path for Dyck paths as well. See also [**Can06**].

Solutions to Exercises

Chapter 1

Solution to Exercise 1.5 Let $\mathrm{frev}(\sigma)$ denote the map on permutations which sends σ to $n - \sigma_n + 1, \ldots, n - \sigma_2 + 1, n - \sigma_1 + 1$. Note that $\mathrm{inv}(\mathrm{frev}(\sigma)) = \mathrm{inv}(\sigma)$. For $\sigma \in S_n$, define

$$(\mathrm{C}.1) \qquad \mathrm{co}\phi(\sigma) = \mathrm{frev}(\phi(\mathrm{frev}(\sigma))).$$

Since the maps ϕ and frev are bijections, so is $\mathrm{co}\phi$. Now

$$(\mathrm{C}.2) \qquad \mathrm{comaj}(\sigma) = \mathrm{maj}(\mathrm{frev}(\sigma))$$

$$(\mathrm{C}.3) \qquad\qquad = \mathrm{inv}(\phi(\mathrm{frev}(\sigma)))$$

$$(\mathrm{C}.4) \qquad\qquad = \mathrm{inv}(\mathrm{co}\phi(\sigma)).$$

Now ϕ fixes the last element, so the last element of $\phi(\mathrm{frev}(\sigma))$ is $n - \sigma_1 + 1$, hence the first element of $\mathrm{co}\phi(\sigma)$ is σ_1. Furthermore, both frev and ϕ fix $\mathrm{Ides}(\sigma)$, hence so does $\mathrm{co}\phi$, i.e. it extends to a map on words of fixed weight. $\square$

The map above can also be described by starting with $\mathrm{co}\phi^{(1)} = \sigma_n$, and recursively defining $\mathrm{co}\phi^{(i)}$ by adding σ_{n-i+1} to the beginning of $\mathrm{co}\phi^{(i-1)}$, drawing bars and creating blocks as before depending on how σ_{n-i+1} and σ_{n-i+2} compare, then cycling by moving the first element of a block to the end, etc.

Solution to Exercise 1.7 The number of squares on or below the diagonal and above the x-axis is $\binom{n+1}{2}$. Now just show that coinv counts squares below the path and above the x-axis, by looking at what happens to area when you interchange consecutive N and E steps. $\square$

Solution to Exercise 1.10

From (1.31),

$$(\mathrm{C}.5) \qquad \frac{(c/a)_n}{(c)_n} a^n =_2 \phi_1 \left(\begin{matrix} a, & q^{-n} \\ & c \end{matrix}; q; q \right) = \sum_{k=0}^{n} \frac{(a)_k (q^{-n})_k}{(q)_k (c)_k} q^k$$

$$= \frac{(a)_n (q^{-n})_n}{(q)_n (c)_n} q^n \sum_{k=0}^{n} \frac{(a)_{n-k}}{(a)_n} \frac{(q^{-n})_{n-k}}{(q^{-n})_n} \frac{(q)_n}{(q)_{n-k}} \frac{(c)_n}{(c)_{n-k}} q^{-k}.$$

Now

$$\frac{(c)_n}{(c)_{n-k}} = (1 - cq^{n-k})(1 - cq^{n-k+1}) \cdots (1 - cq^{n-1})$$

$$= (-cq^{n-k})(1 - c^{-1}q^{-n+k}) \cdots (-cq^{n-1})(1 - c^{-1}q^{-n+1}) \cdots$$

$$= (-c)^k q^{n-k+\ldots+n-1} (c^{-1}q^{1-n})_k.$$

Thus

$$\frac{(a)_{n-k}(c)_n}{(a)_n(c)_{n-k}} = (c/a)^k \frac{(c^{-1}q^{1-n})_k}{(a^{-1}q^{1-n})_k}.$$

Applying this to (c, a) and also to (q, q^{-n}) and plugging into (C.5) and simplifying gives a formula for

$$_2\phi_1 \left(\begin{matrix} c^{-1}q^{1-n}, & q^{-n} \\ & a^{-1}q^{1-n} \end{matrix} ; q; q^n c/a \right).$$

Now replace $a^{-1}q^{1-n}$ by C and $c^{-1}q^{1-n}$ by A to rephrase the result in terms of

(C.6) $$_2\phi_1 \left(\begin{matrix} A, & q^{-n} \\ & C \end{matrix} ; q; q^n C/A \right). \quad \square$$

Solution to Exercise 1.11

After replacing a by q^a, for $|z|$ sufficiently small and $|q| < 1$, it is easy to see the series in (1.23) converges uniformly and hence is analytic, either as a function of z or of q. Thus the limit as $q \to 1^-$ of the series can be taken term by term, which clearly approaches the left-hand-side of (1.33).

To compute the limit as $q \to 1^-$ of $(q^a z)_\infty / (z)_\infty$, take the logarithm and expand using Taylor's series to get

$$\ln \frac{(q^a z)_\infty}{(z)_\infty} = \sum_{i=0}^{\infty} \ln(1 - q^i q^a z) - \ln(1 - q^i z)$$

$$= \sum_{i=0}^{\infty} - \sum_{k=1}^{\infty} \frac{(zq^i)^k}{k}(1 - q^{ak})$$

$$= -\sum_{k=1}^{\infty} \frac{(z)^k}{k} \frac{(1 - q^{ak})}{1 - q^k}$$

$$= -\sum_{k=1}^{\infty} \frac{(z)^k}{k} \frac{(1 - q^{ak})}{1 - q} \frac{1 - q}{1 - q^k},$$

which approaches

$$-\sum_{k=1}^{\infty} \frac{(z)^k}{k} \frac{ak}{k} = -a \ln(1 - z)$$

as $q \to 1^-$. Again uniform convergence can be used to justify the interchange of summation and limits. $\quad \square$

Solution to Exercise 1.14

Just take the logarithm of the product in the generating functions (1.35) and (1.36) and expand using Taylor's series. $\quad \square$

Solution to Exercise 1.18

The truth of (1.42) is fairly easy to show. To obtain (1.41) from it set $x_1 = x_2 = \cdots = x_n = 1$ and $x_i = 0$ for $i > n$. The coefficient of a given t_λ is then $\binom{n}{n_1, n_2, \ldots}$ by the multinomial theorem. Next note that for any $f \in \Lambda$, $f(ps_x^1)$ is a polynomial in x, which follows by expanding f in the p_λ basis and then replacing p_λ by $x^{\ell(\lambda)}$. By (1.41) and the $q = 1$ case of Theorem 1.15, the values of $f(ps_x^1)$ are what we are trying to show they are when $x \in \mathbb{N}$ and $f = e_m, h_m, m_\lambda$. Since two polynomials in x which agree on infinitely many values of x must be identically equal, the result follows. $\square$

Solution to Exercise 1.21

Recall f^λ equals $s_\lambda|_{m_{1^n}}$, the coefficient of the monomial symmetric function m_{1^n} in s_λ. Thus from the Pieri rule (Theorem 1.20 (4)), the sum above equals

$$h_{n-k}h_k|_{m_{1^n}} = < h_{n-k}h_k, h_1^n >$$
$$< h_{n-k}h_k, p_1^n >$$
$$= n!\, h_{n-k}h_k|_{p_1^n},$$

where we have used (1.45) and (1.43). Since by the solution to Exercise 1.14 the coefficient of p_1^k in h_k is $1/k!$, the result follows. $\square$

This can also be done by using the hook formula (1.50) for f^λ, in which case we end up with the sum

$$\sum_{j=0}^{n} \frac{n-2j+1}{n-j+1}\binom{n}{j} = \sum_{j=0}^{n}(1-\frac{j}{n-j+1})\binom{n}{j}$$
$$= \sum_{j=0}^{n}\binom{n}{j} - \sum_{j=1}^{n}\binom{n}{j-1} = \binom{n}{k}.\quad \square$$

Chapter 2

Solution to Exercise 2.5

Let $\epsilon > 0$ be given, and choose δ small enough so that if the mesh of a partition P of the interval $[0, 1]$ is less than δ, then

$$(\mathrm{C.7}) \qquad \left| \int_0^1 f(x)dx - \sum_{i=0}^{N} f(x_i*)(x_{i+1} - x_i) \right| < \epsilon,$$

where $x_i* \in [x_i, x_{i+1}]$. Let $1 - \delta < q < 1$, and let $P = [0, q^N, q^{N-1}, \ldots, q, 1]$. Then

$$(\mathrm{C.8}) \qquad \sum_{i=0}^{N} f(q^i)(q^i - q^{i+1})$$

is a Riemann sum for (f, P), with $x_i* = q^i$. Furthermore, since

$$(\mathrm{C.9}) \qquad q^i - q^{i+1} = q^i(1 - q) < \delta,$$

(C.7) then holds for any N. Thus

$$(\mathrm{C.10}) \qquad \left| \int_0^1 f(x)dx - \sum_{i=0}^{\infty} f(q^i)(q^i - q^{i+1}) \right| \le \epsilon.$$

Letting $\epsilon \to 0$, which forces $q \to 1^-$, we get the result. $\square$

Solution to Exercise 2.7

Since

$$p_k[XY\frac{1-t}{1-q}] = p_k[X]p_k[Y]\frac{1-t^k}{1-q^k},$$

applying $\omega_{q,t}$ to the Y set of variables gives

$$\omega_{q,t} p_k[XY \frac{1-t}{1-q}] = (-1)^{k-1} p_k[X] p_k[Y].$$

It follows that, applying $\omega_{q,t}$ to the Y variables,

$$\omega_{q,t} h_n[XY \frac{1-t^k}{1-q^k}] = e_n[XY].$$

The result now follows by applying it to the other side of (2.12), using (2.16). $\quad\square$

Solution to Exercise 2.10

By the Pieri rule, $e_{n-d} h_d = s_{d+1,1^{n-d-1}} + s_{d,1^{n-d}}$, the sum of two consecutive hook shapes. By (2.24),

$$\begin{aligned}
< \tilde{H}_\mu, e_{n-d} h_d > &= < \tilde{H}_\mu, s_{d+1,1^{n-d-1}} + s_{d,1^{n-d}} > \\
&= e_{n-d-1}[B_\mu - 1] + e_{n-d}[B_\mu - 1] \\
&= e_{n-d}[B_\mu].
\end{aligned}$$

To justify the last line, first note that $1 \in B_\mu$. Thus to take the sum of all possible products of $n - d$ distinct terms in B_μ, we can take either $n - d$ distinct terms from $B_\mu - 1$ to get $e_{n-d}[B_\mu - 1]$, or take $n - d - 1$ distinct terms from $B_\mu - 1$, and also the term 1, to get $e_{n-d-1}[B_\mu - 1]$.

On the other hand, letting $d = 0$ in (2.44) we see

(C.11) $$< \tilde{H}_\mu, e_n >= e_n[B_\mu] = e_{n-1}[B_\mu - 1].$$

Thus (2.24) holds for $k = n - 1$. Next letting $d = 1$ in (2.44) we see

(C.12) $$< \tilde{H}_\mu, s_{1^n} + s_{2,1^{n-2}} > = < \tilde{H}_\mu, s_{1^n} > + < \tilde{H}_\mu, s_{2,1^{n-2}} >= e_{n-1}[B_\mu]$$
$$= e_{n-1}[B_\mu - 1] + e_{n-2}[B_\mu - 1]$$

and so we see (2.24) holds for $k = n - 2$. We continue this way inductively. $\quad\square$

The special cases $d = 0$, $d = n$ imply

$$< \tilde{H}_\mu, h_d > = e_0[B_\mu] = 1,$$
$$< \tilde{H}_\mu, e_n > = e_n[B_\mu] = \prod_{x \in \mu} q^{a'} t^{l'} = t^{n(\mu)} q^{n(\mu')} = T_\mu.$$

Since $V(\mu)^{(0,0)} = \mathbb{C}$, this accounts for the occurrence of the trivial representation. Since the bi-degree of the terms in Δ_μ is T_μ, and since Δ_μ is in $V(\mu)^\epsilon$ by inspection (the diagonal action permutes the columns of the determinant) this accounts for the one and only occurrence of the sign representation. $\quad\square$

Solution to Exercise 2.11

After replacing X by $X/(1 - t)$, then replacing t by $1/t$, we see J_μ gets sent to $t^{-n(\mu)} \tilde{H}_\mu$ and in the right-hand side of (2.46), J_λ gets sent to $t^{-n(\lambda)} \tilde{H}_\lambda$. After some simple rewriting, the problem reduces to showing

(C.13) $$t^{n(\mu)-n(\lambda)} \frac{\prod_{x \in \lambda} t^{l_\lambda + \chi(x \notin B)}}{\prod_{x \in \mu} t^{l_\mu + \chi(x \notin B)}} = 1.$$

Since $\sum_{x \in \mu} l = \sum_{x \in \mu} l' = n(\mu)$, the problem further reduces to showing

(C.14) $$\frac{\prod_{x \in \lambda} t^{\chi(x \notin B)}}{\prod_{x \in \mu} t^{\chi(x \notin B)}} = 1.$$

This is easy, since $x \in \lambda$ and $x \notin B$ implies $x \in \mu$. $\quad \square$

Solution to Exercise 2.12

Assume $f \in DH_n$, and let

$$(C.15) \qquad F = (\prod_{j=1}^{n} \frac{\partial^{a_j}}{x_j^{a_j}} \frac{\partial^{b_j}}{y_i^{b_j}}) f$$

be an arbitrary partial of f. Then

$$(C.16) \qquad \sum_{1=1}^{n} \frac{\partial^h}{x_i^h} \frac{\partial^k}{y_i^k} F = \sum_{1=1}^{n} \frac{\partial^h}{x_i^h} \frac{\partial^k}{y_i^k} (\prod_{j=1}^{n} \frac{\partial^{a_j}}{x_j^{a_j}} \frac{\partial^{b_j}}{y_i^{b_j}}) f$$

$$(C.17) \qquad = \prod_{j=1}^{n} \frac{\partial^{a_j}}{x_j^{a_j}} \frac{\partial^{b_j}}{y_i^{b_j}} \sum_{1=1}^{n} \frac{\partial^h}{x_i^h} \frac{\partial^k}{y_i^k} f$$

$$(C.18) \qquad = \prod_{j=1}^{n} \frac{\partial^{a_j}}{x_j^{a_j}} \frac{\partial^{b_j}}{y_i^{b_j}} 0$$

since $f \in DH_n$. Thus DH_n is closed under partial differentiation. $\quad \square$

To show part (b), let (r, s) be a "corner cell" of μ, so if we remove (r, s) from μ we get the graph of a partition ν say. If we expand Δ_μ along the column corresponding to (r, s) we get

$$(C.19) \qquad \Delta_\mu = \pm \sum_{i=1}^{n} x_i^r y_i^s (-1)^i \Delta_\nu(\hat{X}_i, \hat{Y}_i),$$

where $\hat{X}_i$ is the set of variables X_n with x_i deleted, and similarly for $\hat{Y}_i$. Thus

$$(C.20)$$
$$\sum_{1=1}^{n} \frac{\partial^h}{x_i^h} \frac{\partial^k}{y_i^k} \Delta_\mu = \pm \sum_{j=1}^{n} \frac{\partial^h}{x_j^h} \frac{\partial^k}{y_j^k} \Delta_\mu = \pm \sum_{j=1}^{n} \frac{\partial^h}{x_j^h} \frac{\partial^k}{y_j^k} \sum_{i=1}^{n} x_i^r y_i^s (-1)^i \Delta_\nu(\hat{X}_i, \hat{Y}_i)$$

$$(C.21) \qquad = \sum_{i=1}^{n} x_i^r y_i^s (-1)^i \sum_{\substack{j=1 \\ j \neq i}}^{n} \frac{\partial^h}{x_j^h} \frac{\partial^k}{y_j^k} \Delta_\nu(\hat{X}_i, \hat{Y}_i) + \sum_{i=1}^{n} \Delta_\nu(\hat{X}_i, \hat{Y}_i) \frac{\partial^h}{x_i^h} \frac{\partial^k}{y_i^k} x_i^r y_i^s (-1)^i.$$

By induction on $|\mu|$ we can assume that for fixed i,

$$(C.22) \qquad \sum_{\substack{j=1 \\ j \neq i}}^{n} \frac{\partial^h}{x_j^h} \frac{\partial^k}{y_j^k} \Delta_\nu(\hat{X}_i, \hat{Y}_i) = 0,$$

so the first sum in (C.21) equals zero. The second sum equals

$$(C.23)$$
$$h! \binom{r}{h} k! \binom{s}{k} \sum_{i=1}^{n} \Delta_\nu(\hat{X}_i, \hat{Y}_i) x_i^{r-h} y_i^{s-k} (-1)^i \frac{\partial^h}{x_i^h} \frac{\partial^k}{y_i^k} x_i^r y_i^s (-1)^i = h! \binom{r}{h} k! \binom{s}{k} \det A,$$

where A is the matrix obtained by replacing the column corresponding to the square (r, s) in μ with a column corresponding to $(r - k, s - h)$. Since μ is a partition, A has two equal columns and so $\det A = 0$. $\quad \square$

Chapter 3

Solution to Exercise 3.9

Letting $z = 1/q$ in (3.24), we get

(C.24) $$e_n[X\frac{(1-1/q)}{(1-q)}] = \frac{1-1/q}{1-q}E_{n,1} = \frac{-E_{n,1}}{q},$$

since $(1/q)_k = 0$ for $k > 1$. Now

(C.25) $$e_n[X\frac{(1-1/q)}{(1-q)}] = e_n[-X/q] = \frac{(-1)^n}{q^n}e_n[-\epsilon X] = \frac{(-1)^n}{q^n}h_n[X]$$

by Example 1.26, proving (3.33). $\square$

To show (3.34), we let $E = X/(1-q)$ and $F = 1 - z$ in (1.67) as suggested to get

(C.26)

$$e_n[\frac{X}{1-q}(1-z)] = \sum_{\lambda \vdash n} s_{\lambda'}[\frac{X}{1-q}]s_\lambda[1-z] = \sum_{r=0}^{n-1}(-z)^r(1-z)s_{(r+1,1^{n-r-1})}[\frac{X}{1-q}]$$

using (1.72). Thus

(C.27) $$e_n[\frac{X}{1-q}(1-z)]|_{z^n} = (-1)^n s_{(n)}[\frac{X}{1-q}].$$

On the other hand, taking the coefficient of z^n in the right-hand-side of (3.24),

(C.28) $$\sum_{k=1}^n \frac{(z)_k}{(q)_k}E_{n,k}|_{z^n} = \frac{(z)_n}{(q)_n}E_{n,n}|_{z^n}$$

$$= \frac{(-1)^n q^{\binom{n}{2}}}{(q)_n}E_{n,n}.$$

Comparing coefficients of z^n we get

(C.29) $$E_{n,n} = q^{-\binom{n}{2}}(q)_n h_n[\frac{X}{1-q}] = q^{-\binom{n}{2}}\tilde{H}_{(n)}$$

by (2.43). Now take ∇ of both sides of (C.29). $\square$

Solution to Exercise 3.13

Any path with bounce $= 1$ must begin with $n-1$ N steps followed by an E step. There are exactly $n-1$ of these, with

(C.30) $$\text{area} = \binom{n-1}{2}, \binom{n-1}{2}+1, \ldots, \binom{n-1}{2}+n-2.$$

Thus

(C.31) $$F_n(q,t)|_t = q^{\binom{n-1}{2}}[n-1].$$

Now a path with area $= 1$ must be of the form

(C.32) $$(NE)^j NNEE(NE)^{n-2-j}, \qquad 0 \le j \le n-2,$$

and such a path has bounce $= n-1+n-2+\ldots+2+1-(n-1-j) = \binom{n-1}{2}+j$. Thus

(C.33) $$F_n(t,q)|_t = q^{\binom{n-1}{2}}[n-1]. \quad \square$$

Solution to Exercise 3.18

Given $\pi \in L_{n,n}^+$ having k rows $i_1, i_2, \ldots, i_k$ of length 0, let π' be the path obtained by deleting these k rows of length 0, and also deleting columns $i_1, \ldots, i_k$, and then decreasing the length of each remaining row by 1. If π has r rows of length 1, then π' has r rows of length 0. Clearly

$$(C.34) \qquad\qquad \text{area}(\pi) = n - k + \text{area}(\pi').$$

Now consider the sum of q^{dinv} over all π which get mapped to a fixed π'. Clearly

$$(C.35) \qquad\qquad \text{dinv}(\pi) = \text{dinv}(\pi') + B,$$

where B is the sum of all d-inversion pairs of π which involve pairs of rows both of length 0 or pairs of rows of lengths 0 and 1.

The number of d-inversion pairs involving pairs of rows both of length 0 is simply $\binom{k}{2}$. The number of d-inversion pairs of π involving pairs of rows of lengths 0 and 1 will be the number of inversion pairs in the area sequence $a_1 a_2 \cdots a_n$ of π involving 1's and 0's. When we sum over all possible π with fixed π', the subword of this area sequence consisting of 0's and 1's varies over all multiset permutations of k 0's and r 1's which begin with a 0. The beginning 0 doesn't affect anything, so by MacMahon's result (1.10) the number of inversions of this subword sequence generates a factor of $\begin{bmatrix} k-1+r \\ r \end{bmatrix}$. $\square$

Solution to Exercise 3.19

Call a square w an N-*regular* square of π if w is just to the right of an N step of π. Also, call a square w an E-*regular* square of π if w is just below an E step of π. Note that any row of π contains exactly one N-regular square and every column of π contains exactly one E-regular square. Given a generic column x of π, assume first that in this column the E-regular square w is also N-regular. It is then easy to see that the number of squares of λ satisfying (3.63) in column x equals the number of d-inversion pairs of π involving the row containing w and some row above w.

If w is not also N-regular, travel downwards from w at a 45 angle in a SW direction from w, staying in the same diagonal, until reaching a square z which is N-regular. Note z cannot be E-regular, or the square immediately NE diagonally above it would be N regular which contradicts the choice of z. We claim that the number of squares of λ in column x satisfying (3.63) equals the number of d-inversion pairs of π involving the row containing z and some row above z. The reason is that between the row containing w and the row containing z there cannot be any rows whose length is the same as, or one less than, the length of the z-row. Thus the number of d-inversion pairs involving z and some row above it is the same as if w were an N-regular square.

Thus for each column x of π, we identify an N-regular square z_x of π such that the number of d-inversion pairs involving row z_x and some row above it equals the number of squares satisfying (3.63) in column x. Geometrically, it is easy to see that this gives a bijective pairing between columns and rows, and since dinv clearly equals the sum over all rows r of the number of d-inversion pairs involving row r and rows above it, the result follows. $\square$

Chapter 4

Solution to Exercise 4.4

Changing a D step to a consecutive NE pair leaves the number of lower triangles the same, so (4.8) holds. Now to calculate bounce($\alpha(\pi)$), we first would remove the topmost D step and its row and column, together with the other D steps and their rows and columns, then form the bounce path. Thus the bounce path for $\alpha(\pi)$ is the same as the bounce path for π with the top row and rightmost column truncated. If the bounce path for π intersects the diagonal at k points between $(0,0)$ and (n,n), the contribution of these points to bounce($\alpha(\pi)$) will be reduced by 1 for each point. However, each of the peaks corresponding to each of these k points will have an extra D step above them in $\alpha(\pi)$, so bounce stays the same. □

Solution to Exercise 4.15

By (2.24) we have

$$(C.36) \qquad \left\langle \tilde{H}_\mu, e_{n-d}h_d \right\rangle = e_{n-d}[B_\mu].$$

This, together with the facts that

$$(C.37) \qquad \left\langle \tilde{H}_\mu, s_n \right\rangle = 1$$

for all μ and that the $\tilde{H}_\mu$ form a basis for Λ, implies that for any symmetric function F

$$(C.38) \qquad \langle F, e_{n-d}h_d \rangle = \left\langle \Delta_{e_{n-d}} F, s_n \right\rangle.$$

In particular this holds for $F = \nabla E_{n,k}$. Now apply Theorems 4.14 and 4.47.

To prove the second part of the exercise, note that $S_{n+1,d,1}(q,t) = t^n S_{n,d}(q,t)$, since if there is one step below the lowest E step then it must be a N step, so our path starts with a NE pair which contributes t^n to the bounce statistic. So letting $n = n + 1$ and $k = 1$ in Corollary 4.14.1 yields Corollary 4.14.2.

Solution to Exercise 4.20

Let $(n_0, \ldots, n_{k-1})$ and $(d_0, \ldots, d_k)$ be two N-area and D-area vectors, and consider the sum of q^{dinv} over all π with these N-area and D-area vectors. We construct such a π by starting with a Dyck path $\mathrm{row}_1,,\ldots,\mathrm{row}_{n_0}$ consisting of n_0 N rows of length zero, then we insert d_0 D rows of length zero into this path. The resulting sequence $\mathrm{row}_1, \mathrm{row}_2, \ldots$ corresponds to a path π, and consider the value of $\mathrm{dinv}(\pi)$. Any D row will create inversion pairs with all the N rows before (i.e. below) it, and any pair of N rows will create an inversion pair. Thus as we sum over all ways to insert the D rows we generate a factor of

$$(C.39) \qquad q^{\binom{n_0}{2}} \begin{bmatrix} d_0 + n_0 \\ d_0 \end{bmatrix}.$$

Next we wish to insert the $n_1 + d_1$ rows of length one. For simplicity consider the case where, after inserting these rows, all the N rows of length one occur after (i.e. above) all the D rows of length one. We have the constraint that we cannot insert a row of length one just after a D row of length zero and still have the row sequence of an actual Schröder path. In particular we must have an N row of length zero immediately before the leftmost row of length one. Now each of the rows of length one will create an inversion pair with each N row of length zero before it, but will not create an inversion pair with any of the D rows of length zero. It follows that we can essentially ignore the D rows of length zero, and when summing over all possible insertions we generate a factor of

$$(C.40) \qquad q^{\binom{n_1}{2}} \begin{bmatrix} n_1 + d_1 + n_0 - 1 \\ n_1 + d_1 \end{bmatrix},$$

since each pair of N rows of length one will generate an inversion pair, but none of the D rows of length one will occur in an inversion pair with any row of length one. The factor of $n_0 - 1$ arises since we must have an N row of length 0 before the leftmost row of length 1. In fact, (C.40) gives the (weighted) count of the inversion pairs between rows of length zero and of length one, and between N rows of length one, no matter how the N rows and D rows of length one are interleaved with each other. Thus when we sum over all such possible interleavings of the N and D rows of length one, we generate an extra factor of

$$(C.41) \qquad \begin{bmatrix} n_1 + d_1 \\ n_1 \end{bmatrix}.$$

Thus the total contribution is

$$(C.42) \qquad q^{\binom{n_1}{2}} \begin{bmatrix} n_1 + d_1 + n_0 - 1 \\ n_1 + d_1 \end{bmatrix} \begin{bmatrix} n_1 + d_1 \\ n_1 \end{bmatrix} = q^{\binom{n_1}{2}} \begin{bmatrix} n_1 + d_1 + n_0 - 1 \\ n_1, d_1 \end{bmatrix}.$$

When inserting the rows of length 2, we cannot insert after any row of length 0 and still correspond to a Schröder path. Also, none of the rows of length 2 will create inversion pairs with any row of length 0. Thus by the argument above we get a factor of

$$(C.43) \qquad q^{\binom{n_2}{2}} \begin{bmatrix} n_2 + d_2 + n_1 - 1 \\ n_2, d_2 \end{bmatrix}.$$

It is now clear how the right-hand-side of (4.26) is obtained.

Solution to Exercise 4.23

Begin by replacing n by $n + d$ in Theorem 4.7 to get

$$(C.44) \quad S_{n+d,d}(q,t) = \sum_{b=1}^{n} \sum_{\substack{\alpha_1 + \ldots + \alpha_b = n,\, \alpha_i > 0 \\ \beta_0 + \beta_1 + \ldots + \beta_b = d,\, \beta_i \geq 0}} \begin{bmatrix} \beta_0 + \alpha_1 \\ \beta_0 \end{bmatrix} \begin{bmatrix} \beta_b + \alpha_b - 1 \\ \beta_b \end{bmatrix} q^{\binom{\alpha_1}{2} + \ldots + \binom{\alpha_k}{2}}$$

$$t^{\beta_1 + 2\beta_2 + \ldots + b\beta_b + \alpha_2 + 2\alpha_3 + \ldots + (b-1)\alpha_b} \prod_{i=1}^{b-1} \begin{bmatrix} \beta_i + \alpha_{i+1} + \alpha_i - 1 \\ \beta_i, \alpha_{i+1}, \alpha_i - 1 \end{bmatrix}.$$

Fix a value of $\beta_i \geq 0$ for each $1 \leq i \leq b$. This will correspond to a term in the right-hand-side of (C.44) with $\beta_0 = d - \sum_{i=1}^{b} \beta_i$ as long as d is large enough, in which case the inner summand will involve q-binomial coefficients which depend on the β_i for $i \geq 1$ and the α_i, multiplied by

$$(C.45) \qquad \begin{bmatrix} d - \sum_{i=1}^{b} \beta_i + \alpha_1 \\ \alpha_1 \end{bmatrix},$$

which is the first q-binomial coefficient in the summand. As $d \to \infty$, two things happen. First of all, the number of possible choices for β_i, $1 \leq i \leq b$ increases to eventually include any choice of the $\beta_i \geq 0$. Next, the factor (C.45) multiplying such a term increases monotonically upwards to $1/(q)_{\alpha_1}$. Since everything in sight can be expressed as series of nonnegative terms, as $d \to \infty$ the right-hand-side of (C.44) converges to

(C.46)
$$\sum_{b=1}^{n} \sum_{\substack{\alpha_1+\ldots+\alpha_b=n,\ \alpha_i>0 \\ \beta_1,\ldots,\beta_b \geq 0}} \frac{1}{(q)_{\alpha_1}} \begin{bmatrix} \beta_b + \alpha_b - 1 \\ \beta_b \end{bmatrix} q^{\binom{\alpha_1}{2}+\ldots+\binom{\alpha_k}{2}}$$

$$t^{\beta_1+2\beta_2+\ldots+b\beta_b+\alpha_2+2\alpha_3+\ldots+(b-1)\alpha_b} \prod_{i=1}^{b-1} \begin{bmatrix} \beta_i + \alpha_{i+1} + \alpha_i - 1 \\ \beta_i, \alpha_{i+1}, \alpha_i - 1 \end{bmatrix}.$$

Since each of the β_i are independent of each other, the multisum over $\beta_1, \ldots, \beta_b$ can written as a product of b individual sums. These can be evaluated using form (1.28) of the q-binomial theorem as follows.

(C.47)
$$\sum_{\beta_i=0}^{\infty} t^{i\beta_i} \begin{bmatrix} \beta_i + \alpha_{i+1} + \alpha_i - 1 \\ \beta_i, \alpha_{i+1}, \alpha_i - 1 \end{bmatrix} = \begin{bmatrix} \alpha_{i+1} + \alpha_i - 1 \\ \alpha_{i+1} \end{bmatrix} \sum_{\beta_i=0}^{\infty} t^{i\beta_i} \begin{bmatrix} \beta_i + \alpha_{i+1} + \alpha_i - 1 \\ \beta_i \end{bmatrix}$$

$$= \begin{bmatrix} \alpha_{i+1} + \alpha_i - 1 \\ \alpha_{i+1} \end{bmatrix} \frac{1}{(t^i)_{\alpha_{i+1}+\alpha_i}}. \qquad \square$$

Chapter 5

Solution to Exercise 5.7

From the definition of pmaj, it follows that if P is the primary pmaj parking function for π, then $\mathrm{pmaj}(P) = \mathrm{bounce}(\pi)$. From this and the proof of (5.6), it also follows that the corresponding σ is characterized by the property that if each of the runs of σ are reversed (as in the definition of $\tilde{\sigma}$ from Chapter 6) then σ becomes the reverse permutation $n \cdots 21$. In addition we have that within a given run, the u_i-values are monotone increasing as we move right to left, because if say j and $j+1$ are in the same run, then since car j cannot occur in a column of P to the right of car $j+1$, the increase in area caused by the insertion of car j into the partial parking function sequence associated to P (as in Figure 8 from Chapter 5) is at least one more then the increase in area caused by the insertion of car $j+1$.

Since cars j and $j+1$ are in the same run, they will end up in the same diagonal (rows of the same length) in the parking function Q associated to P as in Figure 8, and moreover car j will be the next car inserted into the partial parking function after inserting car $j+1$. Since the u_i-value associated to car j is greater than the u_i value associated to car $j+1$, the increase in dinv caused by the insertion of car j will be greater than the increase in dinv caused by the insertion of car $j+1$. It follows that car j will be inserted below car $j+1$, and so Q is a maxdinv parking function. $\square$

Solution to Exercise 5.10

The condition on x is equivalent to $x \notin C_n^{n-s_k}(S)$, so the element $z = C_n^{s_k-n}(x)$ is not in S. Thus in the inner sum over T, T can be any subset of $\{1, \ldots, n\}/S$ which includes z, and we can choose the remaining $j-1$ elements of T in $\binom{n-k-1}{j-1}$ ways. $\square$

Solution to Exercise 5.12

By (3.24) and (5.28) we have

$$q^{\binom{k}{2}}t^{n-k}\left\langle \nabla e_{n-k}\left[X\frac{1-q^k}{1-q}\right], h_{1^{n-k}}\right\rangle = q^{\binom{k}{2}}t^{n-k}\left\langle \sum_{j=1}^{n-k}\nabla E_{n-k,j}\begin{bmatrix}k+j-1\\j\end{bmatrix}, h_{1^{n-k}}\right\rangle$$

(C.48)

$$= q^{\binom{k}{2}}t^{n-k}\sum_{j=1}^{n-k}\begin{bmatrix}k+j-1\\j\end{bmatrix}\sum_{\substack{P\in\mathcal{P}_{n-k}\\ \pi(P) \text{ has } j \text{ rows of length } 0}}q^{\mathrm{dinv}(P)}t^{\mathrm{area}(P)}.$$

Now (C.48) can interpreted as the sum of $q^{\mathrm{dinv}(P)}t^{\mathrm{area}(P)}$ over all $P \in \mathcal{P}_n$ which have the cars 1 through k in rows of length 0, in decreasing order when read top to bottom. To see why, note that the factor $q^{\binom{k}{2}}$ accounts for the inversions involving only these cars. By the same reasoning as in HW problem 3.18, if there are j cars in rows of length 1, then they must all be larger than k and so the $\begin{bmatrix}k+j-1\\j\end{bmatrix}$ accounts for the inversions generated by the j cars in rows of length 1 with the cars 1 through k, as we sum over all ways to interleave the j cars with the k cars. The factor t^{n-k} accounts for the change in area as we truncate the rows of length 0.

On the other hand, from the Γ bijection one finds that parking functions with cars 1 through k in column 1 correspond to parking functions with cars 1 through k on the diagonal (in decreasing order).   □

Chapter 6

Solution to Exercise 6.2 Say the ribbons R_1, R_2 of the filling T share k contents in common, and let $C(T)$ be the set of columns which contain either a square of R_1 or a square of R_2 with one of these k contents. Then the set of squares in $T \cup T'$ which are in some column in $C(T)$ form a $2 \times k + 1$ rectangle R, with the upper left-hand square of R in R_1 and the lower right-hand square of R in R_2. It is easy to see that there are two ways to choose T and T' to satisfy these conditions, and $|S(T) - S(T')| = 2$. It follows that $sp(T) - sp(T') = 1$ (and also, that $sp(T) + sp(T') = 1$).   □

Solution to Exercise 6.12

By linearity we can assume that $f = s_\lambda$ for some partition λ. By iterating the Pieri rules from Theorem 1.20 we get

(C.49) $$\langle s_\lambda, h_\mu e_\eta\rangle = \sum_{\beta\subseteq\lambda}K_{\beta,\mu}K_{\lambda'/\beta',\eta}.$$

On the other hand, by Theorem 1.20 (6),

(C.50) $$\omega^W s_\lambda[Z+W]|_{z^\mu w^\eta} = \sum_{\beta\subseteq\lambda}K_{\beta,\mu}K_{\lambda'/\beta',\eta}.$$ □

Solution to Exercise 6.15

By Theorem 4.2 and Corollary 4.19.1, this case of (6.20) is equivalent to a statement involving Schröder paths. To see how, given $\pi \in L^+_{n-d,n-d,d}$, we construct a parking function $R(\pi)$ as follows.

(1) Begin by replacing each D step of π by a consecutive NE pair of steps. Note this keeps area fixed.

(2) Form $R(\pi)$ by placing the cars $n-d+1, \ldots, n-1, n$ just to the right of the d new N steps, and the remaining cars $1, 2, \ldots, n-d$ in the other spots, in such a way that read$(R(P))$ is a $(d), (n-d)$-shuffle. It is easy to see that there is a unique way to do this.

The reader will have no trouble checking that the statistic dinv(π) on Schröder paths from Definition 4.18 equals the statistic dinv$(R(\pi))$ on parking functions, and so the $(d), (n-d)$ case of (6.20) follows from Theorem 4.2.   □

Solution to Exercise 6.19

The q, t-Catalan $C_n(q, t)$ is the coefficient of e_n in ∇e_n, which corresponds to the $\mu = \emptyset, \eta = n$ case of the shuffle conjecture. A given $\sigma \in S_n$ will make a contribution to $C_n(q, t)$ from the fermionic formula (6.37) if and only if $\tilde{\sigma} = N \cdots 21$. Such σ are clearly in bijection with compositions $\alpha_1 + \ldots \alpha_b = n$ of n into b positive integers for some $b > 0$, where α_i is the length of the ith run of σ. Call these "composition permutations". Now if σ_i is in run$_j(\sigma)$, then by definition $w_i(\sigma)$ is the number of elements of σ in run$_j(\sigma)$ which are larger than σ_i, plus the number of elements in run$_{j+1}(\sigma)$ which are smaller then σ_i. For a composition permutation, all the elements in run$_{j+1}(\sigma)$ will be smaller then σ_i. It follows that the values of w_i for i in run$_j$ are

$$(C.51) \qquad \alpha_j - 1 + \alpha_{j+1}, \alpha_j - 2 + \alpha_{j+1}, \ldots, \alpha_{j+1}.$$

For each pair (i, j) we have $B_{i,j} = \emptyset$ and $C_{i,j} = \alpha_j$. The formula from Theorem 3.10 now follows easily.   □

Solution to Exercise 6.29

(1) Since the ith horizontal move of the bounce path is by definition the sum of the previous m vertical moves v_i of the bounce path, the (x_i, y_i) coordinates after the ith horizontal move are

$$(C.52) \quad x_i = m(v_0 + v_1 + \ldots + v_{i-m+1}) + (m-1)v_{m-i+2} + \ldots + 2v_{i-1} + v_i$$
$$y_i = v_0 + v_1 + \ldots + v_i,$$

where $v_i = 0$ for $i < 0$. Clearly $x_i \le my_i$.

(2) By (C.52), if any sequence of $m-1$ vertical steps $v_{i-m+2}, \ldots, v_i$ all equal 0, then (x_i, y_i) is on the line $x = my$ and so the next vertical step v_{i+1} is larger than 0 (or, if $y_i = n$, we are already at (mn, n)). On the other hand, if the path hits the line $x = my$, the next vertical step must be positive (unless we are at (mn, n)). This shows we cannot get an infinite sequence of zero vertical steps, or of zero horizontal steps. By part (a) we are always weakly above the line $x = my$, and if our bounce path ever intersects the line $y = n$ we just head straight over to the point (mn, n). Thus we always terminate.

(3) Let R denote the path consisting of vertical steps $v_0, v_1, \ldots, v_i, \ldots$ and horizontal steps $mv_0, mv_1, \ldots, mv_i, \ldots$. It is easy to see that the area

between R and the line $x = my$ equals $m \sum_i \binom{v_i}{2}$. Furthermore, the region between the bounce path and R is simply a union of rectangles. By (C.52), the rectangle between the ith vertical step of the bounce path and the ith vertical step of R has area $v_i(v_{i-m+2} + 2v_{i-m+3} + \ldots + (m-1)v_i)$. $\square$

Solution to Exercise 6.33

(1) Follows easily from the definitions.
(2) Let w, w' be two squares of an m-path π as in (6.68), and let A and A' be the bottom of the group of m-squares they get magnified into in $P^{(m)}$, respectively. If we denote the diagonals (of slope 1) that A, A' are on in $P^{(m)}$ by $\mathrm{diag}(A)$ and $\mathrm{diag}(A')$, then elementary geometric reasoning shows that

(C.53) $$\mathrm{diag}(A) - \mathrm{diag}(A') = \mathrm{diag}(w) - \mathrm{diag}(w').$$

By the nature of the definition of dinv_m, we can show part (b) holds if the contribution of w, w' to $\mathrm{dinv}_m(P)$ equals the number of d-inversions between the group of m-squares that w gets sent to and the group of m-squares that w' gets sent to, which is a simple calculation. $\square$

Chapter 7

Solution to Exercise 7.4

By definition

(C.54) $$\tau_\epsilon \nabla \tau_1 h_k[\frac{X}{1-q}] = \tau_\epsilon \nabla h_k[\frac{X+1}{1-q}].$$

Using the addition formula (1.66) and recalling that

(C.55) $$\nabla h_j[\frac{X}{1-q}] = q^{\binom{j}{2}} h_j[\frac{X}{1-q}]$$

yields

(C.56) $$\tau_\epsilon \nabla h_k[\frac{X+1}{1-q}] = \tau_\epsilon \sum_{j=0}^{k} q^{\binom{j}{2}} h_j[\frac{X}{1-q}] h_{k-j}[\frac{1}{1-q}]$$

(C.57) $$= \sum_{j=0}^{k} q^{\binom{j}{2}} h_j[\frac{X+\epsilon}{1-q}] h_{k-j}[\frac{1}{1-q}].$$

Again using the addition formula, this last sum equals

(C.58) $$\sum_{j=0}^{k} q^{\binom{j}{2}} \sum_{p=0}^{j} h_p[\frac{X}{1-q}] h_{j-p}[\frac{\epsilon}{1-q}] h_{k-j}[\frac{1}{1-q}]$$

(C.59) $$= \sum_{p=0}^{k} h_p[\frac{X}{1-q}] \sum_{j=p}^{k} q^{\binom{j}{2}} (-1)^{j-p} h_{j-p}[\frac{1}{1-q}] h_{k-j}[\frac{1}{1-q}].   \quad \square$$

Solution to Exercise 7.5

By Theorems 1.15 and 1.27, (7.58) equals

$$(C.60) \qquad \sum_{b=0}^{k-p} q^{\binom{b+p}{2}}(-1)^b h_b\Big[\frac{1}{1-q}\Big] h_{k-p-b}\Big[\frac{1}{1-q}\Big]$$

$$(C.61) \qquad = \sum_{b=0}^{k-p} q^{\binom{b}{2}+\binom{p}{2}+pb}(-1)^b q^{-\binom{b}{2}} e_b\Big[\frac{1}{1-q}\Big] h_{k-p-b}\Big[\frac{1}{1-q}\Big]$$

$$(C.62) \qquad = q^{\binom{p}{2}} \sum_{b=0}^{k-p} q^{pb}(-1)^b e_b\Big[\frac{1}{1-q}\Big] h_{k-p-b}\Big[\frac{1}{1-q}\Big]$$

$$(C.63) \qquad = q^{\binom{p}{2}} \frac{(zq^p;q)_\infty}{(z;q)_\infty}\Big|_{z^{k-p}}$$

$$(C.64) \qquad = q^{\binom{p}{2}} \frac{1}{(z;q)_p}\Big|_{z^{k-p}}$$

$$(C.65) \qquad = q^{\binom{p}{2}} \begin{bmatrix} k-1 \\ k-p \end{bmatrix}. \qquad \square$$

Appendix A

Solution to Exercise A.5

Let

$$(C.66) \qquad G_k(\mu) = \langle \tilde{H}_\mu, h_{n-k} h_1^k \rangle.$$

Then $G_1(\mu) = B_\mu = F_1(\mu)$, so we can assume $k > 1$ and $n > 1$, in which case

$$G_k(\mu) = \langle \tilde{H}_\mu, h_{n-k} h_1^k \rangle = \langle h_1^\perp \tilde{H}_\mu, h_{n-k} h_1^{k-1} \rangle$$

$$(C.67) \qquad = \langle \sum_{\nu \to \mu} c_{\mu,\nu} \tilde{H}_\nu, h_{n-1-(k-1)} h_1^{k-1} \rangle$$

$$(C.68) \qquad = \sum_{\nu \to \mu} c_{\mu,\nu} \langle \tilde{H}_\nu, h_{n-1-(k-1)} h_1^{k-1} \rangle$$

$$(C.69) \qquad = \sum_{\nu \to \mu} c_{\mu,\nu} G_{k-1}(\nu).$$

Since G and F satisfy the same recurrence and initial conditions they are equal.
$\square$

Solution to Exercise A.10

We need to show

$$(C.70) \qquad C_\mu[1-z;q,t] = \prod_{x \in \mu}(1 - zq^{a'}t^{l'}).$$

By letting $X = (z,0,0,\ldots,0)$ and $\alpha = 1/z$ in (A.18) we get

$$(C.71) \qquad C_\mu[1-z;q,t] = \sum_{\tilde{\sigma}:\mu\to\{1,\bar{1}\}} t^{\mathrm{maj}(\tilde{\sigma},\mu)} q^{\mathrm{inv}(\tilde{\sigma},\mu)}(-z)^{\mathrm{neg}(\tilde{\sigma})},$$

where the sum is over all super fillings of μ by 1's and $\bar{1}$'s. Now if $\tilde{\sigma}(s) = \bar{1}$, then clearly s is in the descent set if and only if s is not in the bottom row. One checks that any triple of squares u, v, w, with $w = \text{South}(u)$ and u attacking v, form an inversion triple if and only if $\tilde{\sigma}(v) = \bar{1}$, unless u is in the bottom row, in which case they form an inversion triple if and only if $\tilde{\sigma}(u) = \bar{1}$. It follows that

$$(\text{C}.72) \qquad C_\mu[1 - z; q, t] = \sum_{k=0}^{n} (-z)^n e_k[L_\mu],$$

where L_μ is a certain set of q, t powers associated to each square $s \in \mu$, as on the right in Figure 1. Specifically, for those squares s not in the bottom row we associate $t^{l+1} q^{a'}$, while for those squares in the bottom row we associate q^a. On the left in Figure 1 we have the $q^{a'} t^{l'}$ weights whose sum is $B_\mu(q, t)$ by definition. It is easy to see that $L_\mu = B_\mu(q, t)$ for all μ, proving (C.70). $\square$

FIGURE 1. On the left, the $q^{a'} t^{l'}$ weights whose sum is B_μ, and on the right, the weights whose sum is L_μ.

Solution to Exercise A.18

By (A.51) we have

$$(\text{C}.73) \qquad E_\alpha(X; q, t) = \sum_{\substack{\text{nonattacking fillings } \sigma \text{ of } \alpha'}} z^\sigma q^{\text{maj}(\sigma, \alpha')} t^{\text{coinv}(\sigma, \alpha')}$$

$$\times \prod_{\substack{u \in \mu' \\ \sigma(u) \neq \sigma(\text{South}(u))}} \frac{(1 - t)}{(1 - q^{\text{leg}(u)+1} t^{\text{arm}(u)+1})}.$$

Given σ, we create a new filling $\tau(\sigma)$ of $(\alpha_n - 1, \alpha_1, \ldots, \alpha_{n-1})$ by the following procedure. Keeping the basement fixed, shift all the squares of α' one column to the right, then take the last column, squares $(n + 1, j), 1 \leq j \leq \alpha_n$, and move it to column one. Next, remove square $(1, 1)$ (which contains $\sigma(n, 1)$), and shift the remainder of column one down by one square each. Finally, add one to each entry (again leaving the basement fixed) and replace any $n + 1$'s in the resulting filling by 1's. An example of this procedure is shown in Figure 2.

Since σ is nonattacking we must have $\sigma(n, 1) = n$, hence the factor of x_n on the right-hand side of (A.54). The addition of 1 mod n to each entry, combined with the cyclic shift of columns one to the right, fixes the orientation of each triple. Also, it is easy to see that arm and leg lengths are fixed. It follows that $\text{coinv}(\sigma) = \text{coinv}(\tau(\sigma))$, and furthermore that all of the $(1 - t)/(1 - q^{\text{leg}+1} t^{\text{arm}+1})$ terms are fixed. One thing that does change is squares s in the descent set of σ with $\sigma(s) = n$, and also squares s with $\sigma(s) \neq n$ and $\sigma(\text{South}(s)) = n$, which become

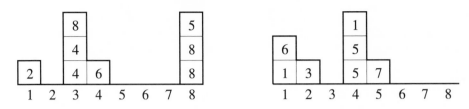

FIGURE 2. A nonattacking filling σ (on the left) and the filling $\tau(\sigma)$ (on the right).

descents in $\tau(\sigma)$ (unless $s = (n, 2)$). A short calculation shows that the change in maj is counterbalanced by the $q^{\alpha_n - 1}$ term, combined with the qx_n substitution in the cyclic shift of variables. □

Appendix B

Solution to Exercise B.4

The portion of $\mathrm{dinv}(A, R)$ in (B.5) coming from (b, c, u, v) quadruples corresponding to pairs of labels in different mutisets is easily seen to equal the LLT $\mathrm{inv}(T(A, R))$ statistic. On the other hand, no pairs of labels within the same multiset L_k are ever counted as LLT inversions, while in $\mathrm{dinv}(A, R)$ they may be counted. If

$$\text{(C.74)} \qquad\qquad a_b^{(u)} - a_c^{(v)} = 0,$$

and the corresponding labels are both in L_k, then the labels must be associated to the same N step, as the labels within L_k correspond to contiguous N steps, and the area of these rows increases by 1 each time as we go up the column. Assume labels $r_b^{(u)}, r_c^{(v)}$ are from paths π_d and π_e, respectively. Then

$$\text{(C.75)} \qquad\qquad r_b^{(u)} < r_c^{(v)} \text{ if and only if } u < v$$

by Condition (2) from Definition B.1. Thus within L_k, any pair of labels associated to a given N step contributes 1 to dinv. Such a pair will end up on the same diagonal of $\beta^{(k)}/\nu^{(k)}$, for a total of $\mathrm{pairdiag}(T)$ such pairs.

On the other hand, if

$$\text{(C.76)} \qquad\qquad a_b^{(u)} - a_c^{(v)} = 1,$$

then since the row lengths increase up the column, within L_k we must have $b = c+1$, and there are no such pairs in (B.5). □

Bibliography

[AAR99]   G. E. Andrews, R. Askey, and R. Roy, *Special functions*, Encyclopedia of Mathematics and its Applications, vol. 71, Cambridge University Press, Cambridge, 1999.

[AKOP02]   G. E. Andrews, C. Krattenthaler, L. Orsina, and P. Papi, *ad-Nilpotent b-ideals in* $\mathrm{sl}(n)$ *having a fixed class of nilpotence: combinatorics and enumeration*, Trans. Amer. Math. Soc. **354** (2002), no. 10, 3835–3853 (electronic).

[And75]   G. E. Andrews, *Identities in combinatorics, II: a q-analog of the Lagrange inversion theorem*, Proc. Amer. Math. Soc. **53** (1975), 240–245.

[Art42]   E. Artin, *Galois theory*, Notre Dame Mathematical Lectures, Notre Dame Ind., 1942, Reprint of the 1976 original.

[Ask80]   R. Askey, *Some basic hypergeometric extensions of integrals of Selberg and Andrews*, SIAM J. Math. Anal. **11** (1980), 938–951.

[Ass07a]   S. Assaf, *A combinatorial proof of LLT and Macdonald positivity*, Preprint, 2007.

[Ass07b]   S. Assaf, *Dual equivalence graphs, ribbon tableaux and Macdonald polynomials*, Ph.D. Thesis, University of California at Berkeley, 2007.

[Ban07]   J. Bandlow, *Combinatorics of Macdonald polynomials and extensions*, Ph.D. Thesis, University of California at San Diego, 2007.

[BBG+99]   François Bergeron, Nantel Bergeron, Adriano M. Garsia, Mark Haiman, and Glenn Tesler, *Lattice diagram polynomials and extended Pieri rules*, Adv. Math. **142** (1999), no. 2, 244–334.

[Ber96]   F. Bergeron, Notes titled "Formularium", 1996.

[BG99]   F. Bergeron and A. M. Garsia, *Science fiction and Macdonald's polynomials*, Algebraic methods and q-special functions (Montréal, QC, 1996), CRM Proc. Lecture Notes, vol. 22, Amer. Math. Soc., Providence, RI, 1999, pp. 1–52.

[BGHT99]   F. Bergeron, A. M. Garsia, M. Haiman, and G. Tesler, *Identities and positivity conjectures for some remarkable operators in the theory of symmetric functions*, Methods Appl. Anal. **6** (1999), 363–420.

[But94]   L. M. Butler, *Subgroup lattices and symmetric functions*, Mem. Amer. Math. Soc. **112** (1994), no. 539, vi+160.

[Can06]   M. Can, *Nabla operator and combinatorial aspects of Atiyah-Bott Lefschetz theorem*, Ph.D. Thesis, University of Pennsylvania, 2006.

[Che95]   Ivan Cherednik, *Nonsymmetric Macdonald polynomials*, Internat. Math. Res. Notices (1995), no. 10, 483–515.

[CL95]   Christophe Carré and Bernard Leclerc, *Splitting the square of a Schur function into its symmetric and antisymmetric parts*, J. Algebraic Combin. **4** (1995), no. 3, 201–231.

[CL06]   M. Can and N. Loehr, *A proof of the q, t-square conjecture*, J. Combin. Theory Ser. A **113** (2006), 1419–1434.

[CR64]   L. Carlitz and J. Riordan, *Two element lattice permutation numbers and their q-generalization*, Duke J. Math. **31** (1964), 371–388.

[EHKK03]   E. Egge, J. Haglund, K. Killpatrick, and D. Kremer, *A Schröder generalization of Haglund's statistic on Catalan paths*, Electron. J. Combin. **10** (2003), Research Paper 16, 21 pp. (electronic).

[ER90]   Ömer Eğecioğlu and Jeffrey B. Remmel, *A combinatorial interpretation of the inverse Kostka matrix*, Linear and Multilinear Algebra **26** (1990), no. 1-2, 59–84.

[FH85]   J. Fürlinger and J. Hofbauer, *q-Catalan numbers*, J. Combin. Theory Ser. A **40** (1985), 248–264.

[Fis95]   S. Fishel, *Statistics for special q, t-Kostka polynomials*, Proc. Amer. Math. Soc. **123** (1995), no. 10, 2961–2969.

[Foa68]   D. Foata, *On the Netto inversion number of a sequence*, Proc. Amer. Math. Soc. **19** (1968), 236–240.

[Fou74]   H. O. Foulkes, *A survey of some combinatorial aspects of symmetric functions*, Permutations (Actes Colloq., Univ. René-Descartes, Paris, 1972), Gauthier-Villars, Paris, 1974, pp. 79–92.

[FS78]   D. Foata and M.-P. Schützenberger, *Major index and inversion number of permutations*, Math. Nachr. **83** (1978), 143–159.

[Gar81]   A. M. Garsia, *A q-analaogue of the Lagrange inversion formula*, Houston J. Math. **7** (1981), 205–237.

[Ges80]   I. Gessel, *A noncommutative generalization and q-analog of the Lagrange inversion formula*, Trans. Amer. Math. Soc. **257** (1980), 455–482.

[GH93]   A. M. Garsia and M. Haiman, *A graded representation model for Macdonald polynomials*, Proc. Nat. Acad. Sci. U.S.A. **90** (1993), no. 8, 3607–3610.

[GH95]   _____, *Factorizations of Pieri rules for Macdonald polynomials*, Discrete Math. **139** (1995), no. 1-3, 219–256,, Formal power series and algebraic combinatorics (Montreal, PQ, 1992).

[GH96a]   _____, *A remarkable q, t-Catalan sequence and q-Lagrange inversion*, J. Algebraic Combin. **5** (1996), no. 3, 191–244.

[GH96b]   A. M. Garsia and M. Haiman, *Some natural bigraded S_n-modules and q, t-Kostka coefficients*, Electron. J. Combin. **3** (1996), no. 2, Research Paper 24, approx. 60 pp. (electronic), The Foata Festschrift.

[GH98]   _____, *A random q, t-hook walk and a sum of Pieri coefficients*, J. Combin. Theory, Ser. A **82** (1998), 74–111.

[GH01]   A. M. Garsia and J. Haglund, *A positivity result in the theory of Macdonald polynomials*, Proc. Nat. Acad. Sci. U.S.A. **98** (2001), 4313–4316.

[GH02]   _____, *A proof of the q, t-Catalan positivity conjecture*, Discrete Math. **256** (2002), 677–717.

[GH06]   I. Grojnowski and M. Haiman, *Affine Hecke algebras and positivity of LLT and Macdonald polynomials*, Preprint, 2006.

[GHT99]   A. M. Garsia, M. Haiman, and G. Tesler, *Explicit plethystic formulas for Macdonald q, t-Kostka coefficients*, Sém. Lothar. Combin. **42** (1999), Art. B42m, 45 pp. (electronic), The Andrews Festschrift (Maratea, 1998).

[GP92]   A. M. Garsia and C. Procesi, *On certain graded S_n-modules and the q-Kostka polynomials*, Adv. Math. **94** (1992), no. 1, 82–138.

[GR98]   A. M. Garsia and J. Remmel, *Plethystic formulas and positivity for q, t-Kostka coefficients*, Mathematical essays in honor of Gian-Carlo Rota (Cambridge, MA, 1996), Progr. Math., vol. 161, Birkhäuser Boston, Boston, MA, 1998, pp. 245–262.

[GR04]   G. Gasper and M. Rahman, *Basic hypergeometric series, second ed.*, Encyclopedia of Mathematics and its Applications, vol. 96, Cambridge University Press, Cambridge, 2004, With a foreword by Richard Askey.

[GT96]   A. M. Garsia and G. Tesler, *Plethystic formulas for Macdonald q, t-Kostka coefficients*, Adv. Math. **123** (1996), 144–222.

[Hag06]   J. Haglund, *The genesis of the Macdonald polynomial statistics*, Sém. Lothar. Combin. **54A** (2005/06), Art. B54Ao, 16 pp. (electronic).

[Hag03]   _____, *Conjectured statistics for the q, t-Catalan numbers*, Adv. Math. **175** (2003), no. 2, 319–334.

[Hag04a]   _____, *A combinatorial model for the Macdonald polynomials*, Proc. Nat. Acad. Sci. U.S.A. **101** (2004), 16127–16131.

[Hag04b]   _____, *A proof of the q, t-Schröder conjecture*, Internat. Math. Res. Notices **11** (2004), 525–560.

[Hai94]   M. Haiman, *Conjectures on the quotient ring by diagonal invariants*, J. Algebraic Combin. **3** (1994), 17–76.

[Hai99]   _____, *Macdonald polynomials and geometry*, New perspectives in algebraic combinatorics (Berkeley, CA, 1996–97), Math. Sci. Res. Inst. Publ., vol. 38, Cambridge Univ. Press, Cambridge, 1999, pp. 207–254.

[Hai00a]   _____, Private communication, 2000.

[Hai00b]   ——, *Private communication*, 2000.

[Hai01]   ——, *Hilbert schemes, polygraphs, and the Macdonald positivity conjecture*, J. Amer. Math. Soc. **14** (2001), 941–1006.

[Hai02]   ——, *Vanishing theorems and character formulas for the Hilbert scheme of points in the plane*, Invent. Math. **149** (2002), 371–407.

[HHL]   J. Haglund, M. Haiman, and N. Loehr, *A combinatorial formula for the nonsymmetric Macdonald polynomials*, Amer. J. Math., to appear.

[HHL05a]   ——, *A combinatorial formula for Macdonald polynomials*, Jour. Amer. Math. Soc. **18** (2005), 735–761.

[HHL05b]   ——, *Combinatorial theory of Macdonald polynomials I: Proof of Haglund's formula*, Proc. Nat. Acad. Sci. U.S.A. **102** (2005), no. 8, 2690–2696.

[HHL+05c]   J. Haglund, M. Haiman, N. Loehr, J. B. Remmel, and A. Ulyanov, *A combinatorial formula for the character of the diagonal coinvariants*, Duke J. Math. **126** (2005), 195–232.

[HL05]   J. Haglund and N. Loehr, *A conjectured combinatorial formula for the Hilbert series for Diagonal Harmonics*, Discrete Math. (Proceedings of the FPSAC 2002 Conference held in Melbourne, Australia) **298** (2005), 189–204.

[Ion03]   Bogdan Ion, *Nonsymmetric Macdonald polynomials and Demazure characters*, Duke Math. J. **116** (2003), no. 2, 299–318.

[Jac70]   H. Jack, *A class of symmetric polynomials with a parameter*, Proc. R. Soc. Edinburgh (A) **69** (1970), 1–18.

[JL01]   G. James and M. Liebeck, *Representations and characters of groups*, second ed., Cambridge University Press, New York, 2001.

[Kad88]   K. Kadell, *A proof of some analogues of Selberg's integral for $k = 1$*, SIAM J. Math. Anal. **19** (1988), 944–968.

[KKR86]   S. V. Kerov, A. N. Kirillov, and N. Yu. Reshetikhin, *Combinatorics, the Bethe ansatz and representations of the symmetric group*, Zap. Nauchn. Sem. Leningrad. Otdel. Mat. Inst. Steklov. (LOMI) **155** (1986), no. Differentsialnaya Geometriya, Gruppy Li i Mekh. VIII, 50–64, 193.

[KN98]   A. N. Kirillov and M. Noumi, *Affine Hecke algebras and raising operators for Macdonald polynomials*, Duke Math. J. (1998), no. 1, 1–39.

[Kno97a]   F. Knop, *Integrality of two variable Kostka functions*, J. Reine Angew. Math. **482** (1997), 177–189.

[Kno97b]   ——, *Symmetric and non-symmetric quantum Capelli polynomials*, Comment. Math. Helv. **72** (1997), 84–100.

[Knu70]   D. E. Knuth, *Permutations, matrices, and generalized Young tableaux*, Pacific J. Math. **34** (1970), 709–727.

[KR86]   A. N. Kirillov and N. Yu. Reshetikhin, *The Bethe ansatz and the combinatorics of Young tableaux*, Zap. Nauchn. Sem. Leningrad. Otdel. Mat. Inst. Steklov. (LOMI) **155** (1986), no. Differentsialnaya Geometriya, Gruppy Li i Mekh. VIII, 65–115, 194.

[KS97]   Friedrich Knop and Siddhartha Sahi, *A recursion and a combinatorial formula for Jack polynomials*, Invent. Math. **128** (1997), no. 1, 9–22.

[KT02]   Masaki Kashiwara and Toshiyuki Tanisaki, *Parabolic Kazhdan-Lusztig polynomials and Schubert varieties*, J. Algebra **249** (2002), no. 2, 306–325.

[Lam]   Thomas Lam, *Schubert polynomials for the affine Grassmannian*, J. Amer. Math. Soc., to appear.

[LLM03]   L. Lapointe, A. Lascoux, and J. Morse, *Tableau atoms and a new Macdonald positivity conjecture*, Duke Math. J. **116** (2003), no. 1, 103–146.

[LLMS06]   Thomas Lam, L. Lapointe, J. Morse, and M. Shimozono, *Affine insertion and Pieri rules for the affine Grassmannian*, preprint, 2006.

[LLT97]   Alain Lascoux, Bernard Leclerc, and Jean-Yves Thibon, *Ribbon tableaux, Hall-Littlewood functions, quantum affine algebras, and unipotent varieties*, J. Math. Phys. **38** (1997), no. 2, 1041–1068.

[LM]   Luc Lapointe and Jennifer Morse, *Quantum cohomolgy and the k-Schur basis*, Trans. Amer. Math. Soc., to appear.

[LM03a]   L. Lapointe and J. Morse, *Schur function analogs for a filtration of the symmetric function space*, J. Combin. Theory Ser. A **101** (2003), no. 2, 191–224.

[LM03b]   ———, *Tableaux statistics for two part Macdonald polynomials*, Algebraic combinatorics and quantum groups, World Sci. Publishing, River Edge, NJ, 2003, pp. 61–84.

[LM03c]   Luc Lapointe and Jennifer Morse, *Schur function identities, their t-analogs, and k-Schur irreducibility*, Adv. Math. **180** (2003), no. 1, 222–247.

[LM04]   L. Lapointe and J. Morse, *Order ideals in weak subposets of Young's lattice and associated unimodality conjectures*, Ann. Comb. **8** (2004), no. 2, 197–219. MR MR2079931 (2005i:06002)

[LM05]   Luc Lapointe and Jennifer Morse, *Tableaux on k + 1-cores, reduced words for affine permutations, and k-Schur expansions*, J. Combin. Theory Ser. A **112** (2005), no. 1, 44–81.

[Loe03]   N. Loehr, *Multivariate analogues of Catalan numbers, parking functions, and their extensions*, Ph.D. Thesis, University of California at San Diego, 2003.

[Loe05a]   Nicholas A. Loehr, *Combinatorics of q, t-parking functions*, Adv. in Appl. Math. **34** (2005), no. 2, 408–425.

[Loe05b]   ———, *Conjectured statistics for the higher q,t-Catalan sequences*, Electron. J. Combin. **12** (2005), Research Paper 9, 54 pp. (electronic).

[LR04]   Nicholas A. Loehr and Jeffrey B. Remmel, *Conjectured combinatorial models for the Hilbert series of generalized diagonal harmonics modules*, Electron. J. Combin. **11** (2004), no. 1, Research Paper 68, 64 pp. (electronic).

[LS78]   A. Lascoux and M.-P. Schützenberger, *Sur une conjecture de H. O. Foulkes*, C. R. Acad. Sci. Paris Sér. A-B **286** (1978), A323–A324.

[LT00]   Bernard Leclerc and Jean-Yves Thibon, *Littlewood-Richardson coefficients and Kazhdan-Lusztig polynomials*, Combinatorial methods in representation theory (Kyoto, 1998), Adv. Stud. Pure Math., vol. 28, Kinokuniya, Tokyo, 2000, pp. 155–220.

[Lus81]   G. Lusztig, *Green polynomials and singularities of unipotent classes*, Adv. in Math. **42** (1981), 169–178.

[LV95a]   L. Lapointe and L. Vinet, *Exact operator solution of the Calogero-Sutherland model*, CRM preprint 2272 (1995).

[LV95b]   ———, *A Rodrigues formula for the Jack polynomials and the Macdonald-Stanley conjecture*, CRM preprint 2294 (1995).

[LV97]   ———, *Rodrigues formulas for the Macdonald polynomials*, Adv. Math. **130** (1997), 261–279.

[LV98]   ———, *A short proof of the integrality of the Macdonald q,t-Kostka coefficients*, Duke Math. J. **91** (1998), 205–214.

[LW]   N. Loehr and G. Warrington, *Nested quantum Dyck paths and $\nabla(s_\lambda)$*, Preprint dated April 25, 2007.

[LW03]   ———, *Private communication*, 2003.

[LW07]   ———, *Square q,t-lattice paths and ∇p_n*, Trans. Amer. Math. Soc. **359** (2007), 649–669.

[Mac60]   P. A. MacMahon, *Combinatory analysis*, Two volumes (bound as one), Chelsea Publishing Co., New York, 1960.

[Mac88]   I. G. Macdonald, *A new class of symmetric polynomials*, Actes du 20e Séminaire Lotharingien, Publ. Inst. Rech. Math. Av. **372** (1988).

[Mac95]   ———, *Symmetric functions and Hall polynomials*, Oxford Mathematical Monographs, second ed., Oxford Science Publications, The Clarendon Press Oxford University Press, New York, 1995.

[Mac96]   I. G. Macdonald, *Affine Hecke algebras and orthogonal polynomials*, Astérisque (1996), no. 237, Exp. No. 797, 4, 189–207, Séminaire Bourbaki, Vol. 1994/95.

[Man01]   Laurent Manivel, *Symmetric functions, Schubert polynomials and degeneracy loci*, SMF/AMS Texts and Monographs, vol. 6, American Mathematical Society, Providence, RI, 2001, Translated from the 1998 French original by John R. Swallow, Cours Spécialisés [Specialized Courses], 3.

[Mar99]   Dan Marshall, *Symmetric and nonsymmetric Macdonald polynomials*, Ann. Comb. **3** (1999), no. 2-4, 385–415, On combinatorics and statistical mechanics.

[Mas]   Sarah Mason, *A decomposition of Schur functions and an analogue of the Robinson-Schensted-Knuth algorithm*, Preprint, 2007.

[Mas06]   S. Mason, *Nonsymmetric Schur functions*, 2006, Ph.D. Thesis, University of Pennsylvania.

[Rei96]   E. Reiner, *A proof of the n! conjecture for generalized hooks*, J. Combin. Theory Ser. A **75** (1996), 1–22.

[Rob38]   G. de B. Robinson, *On the representations of s_n*, Amer. J. Math. **60** (1938), 745–760.

[Sah96]   S. Sahi, *Interpolation, integrality, and a generalization of Macdonald's polynomials*, Internat. Math. Res. Notices (1996), 457–471.

[Sah98]   Siddhartha Sahi, *The binomial formula for nonsymmetric Macdonald polynomials*, Duke Math. J. **94** (1998), no. 3, 465–477.

[Sch61]   C. E. Schensted, *Longest increasing and decreasing subsequences*, Canad. J. Math. (1961), 179–191.

[Sel44]   A. Selberg, *Bemerkninger om et multipelt integral*, Norske Mat. Tidsskr. **26** (1944), 71–78.

[Son04]   C. Song, *The q,t-Schröder polynomial, parking functions and trees*, Ph.D. Thesis, University of Pennsylvania, 2004.

[SSW03]   A. Schilling, M. Shimozono, and D. E. White, *Branching formula for q-Littlewood-Richardson coefficients*, Adv. in Appl. Math. **30** (2003), no. 1-2, 258–272, Formal power series and algebraic combinatorics (Scottsdale, AZ, 2001).

[Sta79]   R. P. Stanley, *Invariants of finite groups and their applications to combinatorics*, Bull. Amer. Math. Soc. (new series) **1** (1979), 475–511.

[Sta86]   ———, *Enumerative combinatorics*, vol. 1, Wadsworth & Brooks/Cole, Monterey, California, 1986.

[Sta88]   D. Stanton, *Recent results for the q-Lagrange inversion formula*, Ramanujan revisited (Urbana-Champaign, Ill., 1987), Academic Press, Boston, MA, 1988, pp. 525–536.

[Sta89]   Dennis Stanton (ed.), *q-series and partitions*, The IMA Volumes in Mathematics and its Applications, vol. 18, New York, Springer-Verlag, 1989.

[Sta99]   R. P. Stanley, *Enumerative combinatorics*, vol. 2, Cambridge University Press, Cambridge, United Kingdom, 1999.

[Sta03]   ———, *Recent progress in algebraic combinatorics*, Bull. Amer. Math. Soc. **40** (2003), 55–68.

[Sta07]   ———, *Catalan addendum*, www-math.mit.edu/~rstan/ec, 2007.

[Ste94]   J. R. Stembridge, *Some particular entries of the two-parameter Kostka matrix*, Proc. Amer. Math. Soc. **121** (1994), 469–490.

[Tes99]   G. Tesler, Private communication, 1999.

[vL00]   Marc A. A. van Leeuwen, *Some bijective correspondences involving domino tableaux*, Electron. J. Combin. **7** (2000), Research Paper 35, 25 pp. (electronic).

[Zab98]   M. Zabrocki, *A Macdonald vertex operator and standard tableaux statistics for the two-column (q,t)-Kostka coefficients*, Electron. J. Combin. **5** (1998), Research Paper 45, 46 pp. (electronic).